Horticulture Business Management

Horticulture Business Management

Editors

Sanjay Sawant, Pandit Patil

scitus
academics

Horticulture Business Management

Edited by **Sanjay Sawant, Pandit Patil**

ISBN: 978-1-68117-070-1

Library of Congress Control Number: 2015935468

SCITUS Academics LLC,
www.scitusacademics.com
Box No. 4766, 616 Corporate Way,
Suite 2, Valley Cottage,
NY 10989

Printed in United States of America on Acid Free Paper ∞

Contents

Preface

Horticulture is the branch of agriculture that deals with the art, science, technology, and business of plant cultivation. It includes the cultivation of fruits, vegetables, nuts, seeds, herbs, sprouts, mushrooms, algae, flowers, seaweeds and non-food crops such as grass and ornamental trees and plants. It also includes plant conservation, landscape restoration, landscape and garden design, construction, and maintenance, and arboriculture. Horticulturists apply their knowledge, skills, and technologies used to grow intensively produced plants for human food and non-food uses and for personal or social needs. Their work involves plant propagation and cultivation with the aim of improving plant growth, yields, quality, nutritional value, and resistance to insects, diseases, and environmental stresses. They work as gardeners, growers, therapists, designers, and technical advisors in the food and non-food sectors of horticulture. Horticulture & Business Management is your stepping stone into this dynamic, innovative and international business! It brings you all the knowledge and skills you need for a bright future in a wonderful green business. The curriculum is comprehensive with both technical and business subjects. You will discover the world of plants, including biology, cultivation techniques and refinement, and learn about the complete chain, from producer to consumer, with subjects that include business and farm economics, management, and marketing.

Editor

Chapter 1

Factors Influencing Willingness to Participate in Multi-Stakeholder Platform by Smallholder Farmers in Northern Ghana: Implication for Research and Development

Edward Martey, Prince M Etwire, Alexander N Wiredu, and Wilson Dogbe

-Savanna Agricultural Research Institute, Nyankpala, Tamale, Ghana

ABSTRACT

Innovation platforms are established to facilitate open communication and collaboration among various actors usually within a value chain to promote collective resource management. The concept of innovation platform as a strategy of entry point for operationalization, validation and dissemination of agricultural technology have received much attention in recent times among researchers in Ghana. However, very limited information exists on the factors that influence willingness of farmers' participation. The present study identified the factors influencing willingness to participate in innovation platform and also tested the level of agreement among the identified constraints associated with participation using the Probit model and Kendall's coefficient of concordance respectively. Data for the study was collected from a cross sectional primary observations of 250 smallholder rice farmers in northern Ghana. The study observes, inter alia, that age of household head, household size and household income significantly influenced willingness to participate on the platform. The Kendall's coefficient of concordance value indicated that there was 21 percent agreement between the rankings of the participation constraints faced by the sampled rice farmers. Income-earning and younger household heads must be targeted coupled with situation of the platforms close to smallholder farmers to encourage participation.

BACKGROUND

Development research in Africa has centered mainly on smallholder commercialization as a way of ensuring food security and economic growth. Agriculture in most African economies is on smallholder basis. In sub-Saharan Africa, Asia and the Pacific, the agriculture-dependent population is over 60%, while in Latin America and high-income economies the proportions are estimated at 18% and 4%, respectively (World Bank [2006]). Most of the smallholder farmers in sub-Saharan Africa are confronted with multifaceted challenges. These problems require several interventions such as institutional reforms that facilitate efficient rural service delivery, development of markets, creation of physical infrastructure, and government policies that are supportive while ensuring a stable and conducive political environment. Smallholder farmers require systems that are responsive to their needs: access to markets, market information, market intelligence and effective farmer organization as the agricultural sector in developing countries transforms towards commercialization.

According to Pingali et al. ([2005]), smallholder farmers in most developing economies find it difficult to participate in markets because of numerous constraints and barriers. These are mostly reflected in the hidden costs that make it difficult to access input and output markets. Transaction costs are the embodiment of access barriers to market participation for most resource-poor smallholders (Delgado [1999]; Holloway et al. [2000]). A fundamental transaction cost these farmers face is the cost of obtaining information (Shepherd [1997]). Though neoclassical economists essentially assume that information is costless, this assumption does not match reality, especially in developing countries (Stiglitz [1988]). The fact that information is not costless has important implications for contracts and transactions, as has been pointed out in work pioneered by Coase ([1937]) and later expanded in Coase ([1960]). Commercialization studies such as Goetz ([1992]), Key et al. ([2000]) and Makhura et al. ([2001]) have identified high transaction costs as one of the key reasons for smallholder farmers' failure to participate in markets. Most are located in remote areas with poor transport, market infrastructure and lack of reliable information on markets and potential exchange partners. In some instances, these transaction costs are so high that markets can be said to be 'missing' (Omamo [1998]; Key et al. [2000]). The adoption and use of multi-stakeholder platform (MSP) in most recent times has the potential to address the high transaction cost incurred by smallholder farmers with regard to market information.

The three most targeted regions in Ghana for most agricultural development projects are Northern, Upper East and Upper West Regions. The regions are inundated with high levels of food insecurity and poverty due to over-reliance on rain-fed agriculture under low farm input conditions. About half of the population of the area face annual food deficit and are net buyers of food (GSS[2008]). Nearly 80 percent of the population depends on subsistence agriculture with very low productivity and low farm income (MoFA [2010]). Several programmes have been designed over the years especially in the rice sector to address the food insecurity and poverty levels in northern Ghana. One of the major interventions was the Multinational NERICA Rice Dissemination Project which sought to contribute to poverty reduction and food security in Ghana through capacity building, technology transfer and production support.

The Commercial Development of Farmer Based Organization (CDFO) of the Millennium Challenge Account (MCA) Programme was an intervention in addressing one of the objectives of the Food and Agriculture Sector Development Policy II (FASDEP II). The objective was to enhance increase competitiveness and integration of farmers into domestic and international markets. To overcome the market constraints of producers, farmers were organized into Farmer-Based Organization (FBO) to access credit and engage in bulk purchase and group marketing to lower the transaction cost they incur in marketing. Agricultural Business Centres (ABCs) were established to enable farmers to access market information on prices, standards and advisory services. The other component of the MCA program sought to strengthen input delivery systems and financial institutions that advances credit to farmers. The intervention however failed to use the integrated approach such as innovation platform to address the marketing problems of the farmers. Most of these interventions fail the test of sustainability after the completion of the entire project.

In an attempt to address the weak linkages among the rice value chain actors as well as increase in rice production in Ghana, the Africa Rice Center (AfricaRice) in collaboration with the Savanna Agricultural Research Institute of the Council for Scientific and Industrial Research (CSIR-SARI) established two Multi-Stakeholder Platforms (MSP) in two rice development hubs[a] in northern Ghana. The Multi-Stakeholder Platform establishment was used to facilitate open communication and collaboration among various actors within the value chain to promote collective resource management. The platform serves as a point of entry, operationalization, validation and dissemination of agricultural technologies. Sustainable development of the rice sector and the rice value chain through an MSP approach prompts key stakeholders to identify and prioritize their needs, concerns, constraints, and vision for key activities and interventions. They also act as a governing entity that coordinates decision-making at local, regional, and national levels regarding rice sector development in the long run. The MSP was also intended to assist farmers overcome their production, marketing and post-production challenges in the regions.

[a]Hubs represent key rice ecologies and different market opportunities across African countries, linked to major national or regional rice-development efforts to facilitate broader uptake of rice knowledge and technologies. It involves large groups of farmers and other value chain

actors such as rice millers, input dealers and traders.

In northern Ghana, the MSP for the rice sector consists of researchers, producers, processors, traders, financial institutions, input dealers, tractor operators and policy makers. The platform brings together all the necessary stakeholders of the rice sector to dialogue on ways of increasing rice production to enhance food security with Savanna Agricultural Research Institute providing the backstopping. Introduction, operationalization, validation and dissemination of all rice development programmes in the region occurs on the platform. All the various actors play a significant role in ensuring a boost in the rice sector as well as strengthening of the rice value chain. Most established platforms are confronted with some challenges which include meeting and sustaining participants' interest, effective participation and collaboration, conflict resolution and collective action.

Problem Statement and Justification

Despite the benefits of the platform especially in transforming the rice value chain system, it is critical to identify the potential factors that are likely to influence farmers' willingness to participate in such a platform as well as the constraints associated with participation. Currently, there is no documented evidence of the factors and constraints limiting farmers' willingness to participate on innovative platforms especially in the study area. It is against this backdrop that the present study attempts to fill the gap in knowledge. The documented factors and constraints will serve as the basis for addressing the sustainability issues regarding the intervention. The studies contribute significantly to the scanty literature on MSP and also serve as a guide for future implementation of similar platforms in other parts of Africa. Finally, facilitators of MSP and rice development programmers will benefit from the knowledge generated by the study.

LITERATURE REVIEW

Concept and Empirical Evidence of Multi-Stakeholder Platforms

The term multi-stakeholder platform (MSPs) describes processes which aim to bring together all major stakeholders in a new form of communication, decision-finding (and possibly decision-making) on a particular issue. They are also based on recognition of the importance of achieving equity and accountability in communication between stakeholders, involving equitable representation of three or more stakeholder groups and their views. They are based on democratic principles of transparency and participation, and aim to develop partnerships and strengthened networks between stakeholders. MSPs cover a wide spectrum of structures and levels of engagement. They can comprise dialogues on policy or grow into consensus-building, decision-making and implementation of practical solutions. The exact nature of any such process will depend on the issues, its objectives, participants, scope, time lines, etc. (www.earthsummit2002.org)

Multi-stakeholder platforms were first proposed in the context of natural resource management, where stakeholders share a common-pool resource, such as access to water in a river basin, and the platform contributes to the collective management of the resource (Rölling et al. [2002]). A platform has value for stakeholders, because they are, or may become, interdependent. Interdependence can create tension, conflict, maneuvering to seek advantage and even group displacement. But it also opens opportunities for mutual understanding, building confidence, social learning and joint action (Rölling et al. [2002]). The platform makes possible actions that none of the members could have achieved on their own. Because of its complex membership and potential for conflict, a platform is likely to require facilitation and may have a lengthy initial phase of mutual learning and role definition, before it can get down to business (Thiele et al. [2005]).

Most value chain does not frequently use the approach of platforms. Several studies including a recent overview of collective action for small farmer market access considered small farmer organizations without mentioning platforms (Markelova et al. [2009]). An exception is the study by Vellema et al. ([2009]), which analyses the oil seed sub-sectoral platform in Uganda. A platform can perform three different, but interlinked, functions in a value chain. First, it can create a space for learning and joint innovation, as an innovation intermediary or broker. Second, it can perform a governance function within the value chain to improve coordination of business activities by actors and reduce transaction costs. Finally, a platform can perform advocacy functions to secure policy change or influence.

Value chain governance may be provided by: (a) market mechanisms, (b) hierarchical non-market mechanisms and (c) by non-market-based voluntary coordination between actors of a collective action type (Markelova et al. [2009]). Dorward et al. ([2009]), writing from a New Institutional Economics perspective, noted that coordination provided through different non-market mechanisms can help market actors reduce transaction costs and escape the low-level equilibrium trap associated with underdeveloped economies, as a weak institutional environment and high transaction risks limit investment opportunities.

Countries that have witnessed success in overcoming marketing constraints by using the MSP include Peru, Bolivia and Ecuador. The platform Cadinas Agricolas Productivas de Calidad (CAPAC) in Peru which focused on native potatoes grown largely by poorer farmers primarily for home consumption or local markets has been instrumental in providing transparent information on price and volumes. The CAPAC has also assumed an important role in market governance by linking farmers' native potato production to Frito-Lay and Ayllin Papa through intermediary NGOs (Thiele et al. [2011]). The platform Andino Boliviana in Bolivia played an important facilitating and coordinating role so as to link farmers' organizations with exporting companies, contributing to the supply of quality chun˜o for the export market (Thiele et al. [2011]). In Ecuador, Chimborazo platform played the most extensive role in market governance. The platform developed and monitored production plans with farmer quotas by market catchment areas to smooth the supply of potatoes to clients. In addition, the platform empowered farmers' organizations and associations to assume a greater leadership role which led to the creation of CONPAPA (Consorcio de Pequen˜os Productores de Papa, Consortium of Small Potato Producers), which from late 2006 took over the technical assistance functions, production planning, bulking up and marketing functions that the Chimborazo platform had previously performed (Thiele et al. [2011]).

Evidence from the Innovation Platform for Technology Adoption (IPTA) in Burkina Faso indicates that maize producers and processors are making profit on their investments and processors are making a difference of only CFA 1.00 over and above the profit margin of producers. Farmers are also increasing the area under maize cultivation due to a guaranteed market as provided by the platform.

Most of the literature reviewed on MSP is descriptive in presentation with little information on specific quantifiable factors that influence willingness of participation. However, the current study used a regression model to analyze farmers' willingness to participate in MSP by drawing lessons from the reviewed literature. Future studies can look at the extent of participation in MSP on household income and food security.

METHODS

Study Area

Northern Ghana consists of Northern, Upper East and Upper West Regions. The vegetation is a typical Guinea Savannah type characterized by drought-resistant grasses and trees. The area plays an important role in agriculture and is normally referred to as the grain basket of the country. It shares borders with Republic of Togo to the east, Ivory Coast to the west and Burkina Faso to the north. Within the country, the northern Ghana is bordered by Volta region to south east and Brong-Ahafo region to the south east. Geographically, the three regions are between longitude 8°46′01.88″ N and 10°58′34″ N and latitude 2°45′45.40″ W and 0°32′59.95″ E and covers a total land area of 97,666 km^2 with an estimated population of 3,317,478. The major crops cultivated in the area are rice, maize, sorghum, millet, groundnut, beans and Shea butter. Livestock and poultry production are also important ventures for most households. The study area consists of two main rice development hubs namely the Navrongo and Savelegu hubs.

Data and Sampling Technique

The study was conducted between August and September, 2012. The basic information for the analysis was obtained from primary data collected with the aid of a structured questionnaire. A total of 250 smallholder rice producers were systematically selected and interviewed. The selection of the rice producers followed a multi-stage systematic random sampling technique. In the first stage, two (2) rice hubs in Navrongo and Savelugu districts were purposively selected. Secondly, five (5) communities each were randomly selected from a list of rice producing communities in each of the rice hubs. Within the selected communities in each of the hubs, 26 and 24 rice producers were further selected from a list of rice producers in the Navrongo and Savelugu selected communities respectively (Table 1). In addition to the survey, key informants interviews and focus group discussion were conducted to augment the household survey.

Table 1: Sampling frame

Rice hub	Communities	Households
Navrongo	5	130
Savelegu	5	120
Total	10	250

Martey *et al.*

Martey *et al. Agricultural and Food Economics* 2014 **2:**11 doi:10.1186/s40100-014-0011-4

Probit Model

The probit regression model was employed to quantify the factors that determine the willingness of farmers to participate in multi-stakeholder platform due to the dichotomous nature of the dependent variable. The justification for the use of the probit model over the logit model is as a result of its ability to constrain the utility value of the decision to join variable to lie within 0 and 1, and its ability to resolve the problem of heteroscedasticity (Asante et al. [2011]). Willingness to participate in MSP (Y) was captured as a dummy variable with the value of 1 assigned to a farmer who is willing to participate and 0 for otherwise. Following from Greene ([2003]), the binary probit for the two choice models can be written as;

$$Y_i^* = \begin{cases} 1 \, if \; Y_i^* > Y \\ 1 \, if \; Y_i^* \leq Y \end{cases} \tag{1}$$

The probit model is given by:

$$P\left(Y = \frac{1}{X}\right) = F(XB) = \frac{1}{\sqrt{2\pi}} \int_{-\infty}^{XB} e^{\frac{-(XB)^2}{2}} dx \tag{2}$$

Where:

$$X = (1, x_{1i}, x_{2i} \ldots\ldots\ldots\ldots, x_{ki}) \tag{3}$$

$$\beta' = (\beta_0, \beta_1, \ldots\ldots\ldots\ldots\ldots, \beta_k) \tag{4}$$

Specifically, the empirical model for determining the willingness to participate in in MSP is specified in equation as follows (5).

$$\begin{aligned} Y_i = \beta_0 + \beta_1 Age + \beta_2 Gend + \beta_3 Mari + \beta_4 HSize + \beta_5 Edu + \beta_6 Assoc \\ \beta_7 Lab + \beta_8 FarmDec + \beta_9 Dist + \beta_{10} Inc + \mu_i \end{aligned} \tag{5}$$

Where Y_i and μ_i represents willingness to participate in MSP and the error term respectively. The set of potential explanatory variables, definitions and their *a priori* expectations are presented in Table 2.

Table 2: Explanatory variables for the probit model

Variable	Definition	Expected sign
Gender (Gend)	1 if male and 0 otherwise	+/–

Age (Age)	Age of household head in years	-
Marital status (Mar)	1 if married and 0 otherwise	+
Education (Edu)	Number of years of formal education	+
Household size (HSize.)	Number of household members that assist with farming	+
Land availability (Lab)	Total land size cultivated in 2012	+
Total income (Inc.)	Total income earned from rice sales in 2012	+
Membership of association (Assoc.)	1 if farmer belongs to FBO and 0 otherwise	+
Distance (Dist.)	Distance from farmers house to the meeting place in km	-
Major farming decision (FarmDec)	1 if household head makes decision alone and 0 otherwise	+/–

Martey *et al.*

Martey *et al. Agricultural and Food Economics* 2014 **2**:11 doi: 10.1186/s40100-014-0011-4

The marginal effect of the variables is calculated using the formula:

Marginal effects=$B_i Ø(Z)$where B_i are the coefficient of the variable (6)

Where B_i represents the estimated coefficients and the Ø (Z) is the cumulative normal distribution value associated with the mean dependent variable from the Probit estimation.

Table 2 shows the set of potential explanatory variables that are expected to explain variation in willingness to participate in MSP and their *a priori* expectations.

Description of Explanatory Variables

Gender is expected to have a significant effect on participation in MSP. Males are expected to have higher willingness of participation in MSP relative to females. Female farmers usually lack access to agricultural resource that enhances their participation in social and innovation platforms. Age is expected to influence willingness to participate in MSP negatively. It is a proxy measure of experience and availability of resources. According to Etwire et al. ([2013]), younger farmers are usually innovative, risk loving and may want to try new concepts. Marital status allows an individual to have access to information and resource. It is expected that household head that are married are more willing to participate on the platform. Nnadi and Akwiwu ([2008]) noted that marriage increases a farmer's concern for household welfare and food security which is therefore likely to have a positive effect on their decision to participate in an agricultural project.

Education is posited to have a positive effect on willingness to participate on the platform. Education enables an individual to make independent choices and to act on the basis of the decision, as well as increase the tendency to co-operate with other people and participate in group activities (Enete and Igbokwe [2009]; South worth and Johnston [1967]; Schultz [1945] and Ofori [1973]). It is also possible that education could increase the chances of the household head earning non-farm income. This could reduce the household dependency on agriculture and thus participation on the platform. Household size is expected to positively influence farmers' willingness to participate on the platform. Household size serves as a form of family labour and compliments the effort of the household heads on the farm (Martey et al. [2013]). The availability of family labour provides the household head the opportunity to share responsibility and save time for other useful activities like the MSP. Also, larger households spend more on food and other household needs. The higher expenditure associated with larger households sizes tend to make them more resource constrained and hence the need for external support (Martey et al. [2013]).

Land availability has an indirect positive effect on farmers' willingness to participate. Household heads with more land are able to produce more *ceteris paribus* and earn higher income to overcome the transportation constraint as well as financial demands of the platform. Most researchers have found a positive relationship between farm size and decision to join or adopt (Adimado [2001]; Kheralla et al.[2001]; Langyintuo and Mekuria [2005]). Other studies such as Mussei et al. ([2001]), and Gockowski and Ndoumbe ([2004]) found a negative relationship between farm size and decision to join or adopt. The coefficient of this variable can be either positive or negative. A household head that earns higher income from rice sales will be more willing to participate on the platform. Higher income household heads are able to meet the financial demands of any group they belong to relative to lower income households. Household head membership of an association/group increases access to information which is important to production and marketing decisions (Olwande and Mathenge[2010]). Most farmer groups engage in group marketing, bulk purchasing of inputs and credit provision for its members. It is therefore expected that household head membership of association/ group will positively affect willingness to participate on the platform. Distance serves as a major cost constraint to willingness to participate on the platform. Household heads residing farther away from the meeting place may find it difficult to effectively participate especially in situations where they do not have means of transport.

The Kendall's Coefficient of Concordance

The Kendall's concordance analysis was used to test for the agreement between the ranked constraints of willingness to participate on the MSP. It establishes the extent of disagreements and agreements among responses. The Kendall's coefficient of concordance (W) is the measure of the degree of agreement among *m* set of *n* ranks. *W* is an index that measures the ratio of the observed variance of the sum of ranks to the maximum possible variance of sum of ranks. If the rankings are in perfect agreement, the variability among sums will be a maximum (Mattson [1986]). Computing the total rank score for each constraint and objective, the constraint and objective with the least score is ranked as the most pressing whilst the one with the highest score is ranked as the least pressing. The total rank score computed is then used to calculate for the coefficient of concordance (*W*), to measure the degree of agreement in the rankings (Edwards [1964]). The formula for the coefficient of concordance W is then given by:

$$W = \frac{\left(\sum T^2 - (\Sigma T)^2 / n\right)/n}{m^2(n^2 - 1)/12} \tag{7}$$

The formula is further simplified as follows:

$$W = \frac{12\left[\sum T^2 - (\Sigma T)^2 / n\right]}{nm^2(n^2 - 1)} \quad (8)$$

Where;

T = sum of ranks for each thing being ranked.

m = number of rankings (smallholder farmers) and

n = number of things (factors) being ranked

The following hypothesis was tested for the potential constraints as follows:

H_0: There is no agreement among the constraints faced by smallholder rice farmers' willingness to participate on the MSP.

H_1: There is agreement among the constraints faced by smallholder rice farmers' willingness to participate on the MSP.

The Coefficient of concordance *W* was tested for significance using the F distribution.

RESULTS AND DISCUSSION

Multi-Stakeholder Platform Participation by Smallholder Farmers

Participation in multi-stakeholder platform based on gender revealed that 59% of the sampled male-headed households whereas 43% of the female-headed households were members on the platform (Table 3). The higher representation by males relative to females on the platform was as a result of unlimited access to agricultural resources that enhances participation in social platforms. Easy access to agricultural resource by male-headed households makes them more adaptive to new innovations and technology compared to female-headed households.

Table 3: Membership of MSP by gender

Membership of MSP	Gender of household head					
	Male	Percentage	Female	Percentage	Total	Percentage
No	61	41.50	59	57.30	120	48.00
Yes	86	58.50	44	42.70	130	52.00
Total	147	100.00	103	100.00	250	100.00

Source: Author's Household Survey Data (2012).

Martey *et al.*

Martey *et al. Agricultural and Food Economics* 2014 2:11 doi:10.1186/s40100-014-0011-4

Willingness to participate in MSP based on gender revealed that 82% (120) of male-headed households were willing to participate on the platform. Female headed households were more willing to participate on the platform as represented by 92% of the sampled farmers (Table 4). In northern Ghana, it is mostly common to see women group flourishing. Female-headed households are more likely to adopt a technology or innovation that will enhance their productivity and income. This finding implies that gender must be critically considered in the selection of the participants. Marketing of agricultural produce is mostly dominated by women who are also faced with myriads of marketing constraints such as lower market price and higher transaction cost. The opportunity offered by the platform as a way of overcoming these constraints as well as knowledge sharing could explain for the higher willingness of participation on the part of female producers.

Table 4: Willingness to participate in MSP by gender

MSP participation	Gender of household head					
	Male	Percentage	Female	Percentage	Total	Percentage
Yes	120	81.60	95	92.20	215	86.00
No	27	18.40	8	7.80	35	14.00
Total	147	100.00	103	100.00	250	100.00

Source: Author's Household Survey Data (2012).

Martey *et al.*

Martey *et al. Agricultural and Food Economics* 2014 2:11 doi:10.1186/s40100-014-0011-4

Table 5 shows that 52% (92) of the sampled households who are members of the MSP are within the age bracket of less than or equal to 40 years (≤40) whereas 46% and 73% of household head falls within the age category of 41 to 60 years and greater than 60 years (>60 years) respectively. In absolute terms majority of the members of the MSP are within the economically active age group of the country. The result has implication on technology adoption and sustainability. Relatively younger household heads should be targeted for the innovation platform approach to enhancing food security and reduction in poverty. Younger household heads are more likely to adopt technology and innovations relative to older household heads (Enete and Igbokwe [2009]).

Table 5: Membership of MSP by age

Membership of MSP	Household head age (Years)							
	≤40	%	41-60	%	> 60	%	Total	%
No	84	47.70	32	54.20	4	26.70	120	48.00
Yes	92	52.30	27	45.80	11	73.30	130	52.00
Total	176	100.00	59	100.00	15	100.00	250	100.00

Source: Author's Household Survey Data (2012).

Martey *et al.*

Martey *et al. Agricultural and Food Economics* 2014 **2**:11 doi:10.1186/s40100-014-0011-4

The results in Table 6 also show high level of willingness of participation across all the age categories. The willingness of participation response is relatively higher among the economically active age group (41–60 years) and older age category (>60 years) relative to the younger age group category (≤40 years) (Table 6). Willingness to participate on the MSP was highly correlated with age where the older age category (>60 years) recorded the highest percentage. The result implies that there is opportunity to reach out to other smallholder farmers irrespective of the age. For sustainability of the platform, it is more expedient for the facilitators to target both the young and the economically active age group category.

Table 6: Willingness of participation based on age

MSP participation	Household head age (Years)							
	≤40	%	41-60	%	> 60	%	Total	%
Yes	149	84.70	52	88.10	14	93.30	215	86.00
No	27	15.30	7	11.90	1	6.70	35	14.00
Total	176	100.00	59	100.00	15	100.00	250	100.00

Source: Author's Household Survey Data (2012).

Martey *et al.*

Martey *et al. Agricultural and Food Economics* 2014 **2**:11 doi:10.1186/s40100-014-0011-4

Fifty-five (55) percent of the household heads belonging to a Farmer-Based Organization (FBO) were members of the MSP whilst 50% of non-members of FBO household head were also members of the platform (Table 7). Farmer Based Organization is one of the main channels for reaching out to numerous farmers for collective actions. It also serves as a platform for capacity building, marketing, bulk purchasing and advocacy. Increase in MSP participation by members of FBO can only be realized through consistent education of farmers on the concept of the MSP.

Table 7: Membership of MSP by membership of FBO

Membership of MSP	Membership of FBO					
	Yes	Percentage	No	Percentage	Total	Percentage
No	51	45.50	69	50.00	120	48.00
Yes	61	54.50	69	50.00	130	52.00
Total	112	100.00	138	100.00	250	100.00

Source: Author's Household Survey Data (2012).

Martey *et al.*

Martey *et al. Agricultural and Food Economics* 2014 **2**:11 doi:10.1186/s40100-014-0011-4

Members and non-members of FBO were more willing to participate in the MSP (Table 8). About 87% of household heads belonging to a FBO were willing to participate in the MSP whilst 86% of non-members of FBO were also willing to join the platform. Previous benefits from

FBOs and the need for social networks may have influenced the willingness decision of household heads in northern Ghana.

Table 8: Willingness of participation in MSP by membership of FBO

MSP participation	Membership of FBO					
	Yes	Percentage	No	Percentage	Total	Percentage
Yes	97	86.60	118	85.50	215	86.00
No	15	13.40	20	14.50	35	14.00
Total	112	100.00	138	100.00	250	100.00

Source: Author's Household Survey Data (2012).

Martey *et al.*

Martey *et al. Agricultural and Food Economics* 2014 **2**:11 doi:10.1186/s40100-014-0011-4

Determinants of Multi-Stakeholder Platform Participation

The probit model was used to estimate the parameters of the determinants of willingness to participate on the multi-stakeholder platform by smallholder farmers in northern Ghana. The E-views software was used to estimate these parameters as well as the marginal effects. The McFadden R-squared value indicates that 36 percent of the variation in the willingness to participate on the platform was explained by the independent variables. The significant likelihood ratio (LR) of 73.43 with 10 degrees of freedom indicates that at least one of the variables in the model had a significant effect on farmers' willingness to participate on the platform and that the explanatory variables jointly influence the farmers' willingness of participation on the platform (Table 9). Farmers' willingness to participate on the multi-stakeholder platform was significantly determined by age of the household head, household size and household income. Numerically and statistically, age was the most influential determinant of willingness to participate in MSP.

Table 9: Determinants of willingness of MSP participation

Variable	Estimated result of probit model		
	Coefficient	Std error	Marginal effect
Age of household head	−0.0688	0.0106	−0.0757***
Gender	0.0001	0.2592	0.0002
Marital status of household	0.0185	0.6966	0.0203
Household size	0.0302	0.0174	0.0332*
Years of education of household head	0.0616	0.0463	0.0678
Membership of association	0.3354	0.2533	0.3689
Land availability	−0.0802	0.0561	−0.0882
Major farming decision	−0.0628	0.3570	−0.0691
Distance to meeting place	−0.0456	0.0663	−0.0501
Household income	0.0004	0.0002	0.0004**
Constant	3.5952	0.9925	3.9547
Number of Observations	250	Log likelihood	−64.5260
Mean dependent var.	0.8600	McFadden R-squared	0.3626

Source: Regression Estimation from Author's Household Survey Data (2012) ***p<0.01, **p<0.05 and *p<0.10.

Martey *et al.*

Martey *et al. Agricultural and Food Economics* 2014 **2**:11 doi:10.1186/s40100-014-0011-4

Age of the household head was significantly associated with a lower probability of willingness to participate on the platform. The probability of willingness to participate in MSP decreases by 7.6 percent for every additional year added to the age of the household head (Table 9). It can be inferred from the result that younger household heads were more willing to participate on the platform than older household heads. Older household heads are normally experienced in farming and may have more social networks which enhance their farming business thus will be more reluctant in joining the platform. Secondly, older household heads may not have the motivation to join such a platform as the level of willingness to adopt innovation declines with age. The result is consistent with Ayamga ([2006]) who found that as age increases, the probability of a farmer to participate in microcredit programmes in northern Ghana, decreases. However, the finding is contrary to Asante et al. ([2011]). They established a positive relationship between age and farmers' decision to join farmer based organization in Ghana.

Household size was significantly associated with a higher probability of farmer's willingness to participate on the platform. The probability of farmer's willingness to participate increases by 3.3 percent for every additional member added to the household (Table 9). Household size represents the supply of family labour for production activities. Farming in the tropics is mostly rain-fed and labour-intensive. A household head with large household size will be more willing to participate on the platform because of excess or additional labour to work on the farm whilst absent from the farm. The demand of economically inactive household members coerces household heads to search for innovative ways to improve upon their farming business.

Willingness to participate on the platform was significantly influenced by household income. The likelihood of farmers' willingness to participate on the platform increases by 0.04 percent for every additional increase in household income (Table 9). Transportation cost is one of the major constraints in MSP participation. Household head with higher income are able to overcome this cost and also make financial contributions in the form of dues and levy as demanded by the platform to ensure its sustainability. Household heads with higher income may necessarily participate on the platform to widen their social networks. Asante et al. ([2011]) also established a positive relationship between farmers' income and willingness to join Farmer Based Organizations. According to their findings, increasing farmers' income by one Ghana cedi increases the likelihood of joining FBOs by 0.026%.

Constraints in Multi-Stakeholder Platform Participation

The constraints of smallholder rice farmers identified during the survey are presented in Table 10. The Kendall's 'W' was found to be 0.205 and significant at 1% level. The null hypothesis was rejected in favour of the alternate hypothesis which stated that there is agreement among the constraints faced by smallholder farmers in MSP participation. The

Kendall's 'W' of 0.205 indicates that there was 21 percent agreement between the respondents in the ranking of the constraints faced by smallholder rice farmers with respect to participation on the platform (Table 10).

Table 10: Constraints in multi-stakeholder platform participation

Identified constraints	Mean rank
Distance to meeting place	3.14
Confliction of meeting days with market days	3.18
Information dissemination regarding meeting days	3.70
Risk	3.77
Weather condition	3.82
Means of transport	4.38
Farming season for holding meeting	6.00
Number of observation	250
Kendall's W	0.205
Chi-square	307.110
Df.	6
Assymp. Sig.	0.000

Source: Estimation from Author's Household Survey Data (2012).

Martey *et al.*

Martey *et al. Agricultural and Food Economics* 2014 **2**:11 doi:10.1186/s40100-014-0011-4

Among the identified ranked constraints, distance to the meeting place, confliction of meeting days with market days, poor dissemination of information regarding meeting days and risk were the top four most constraining factors of farmers' willingness to the MSP participation. Distance to the meeting place was found to be the most constraining factor of MSP participation. The success of the MSP is highly dependent on the location of the meeting place. The situation becomes aggravated where there is no access to transport services. Conflicting meeting days with market days was another major constraint to a successful willingness to participate on the platform. Most of the smallholder rice farmers are engaged in marketing activities either as sellers or buyers. Marketing of farm produce generates income for the household.

Timely dissemination of information about meeting days was crucial to ensuring maximum participation on the platform. Season for holding meeting was the least identified constraint according to the respondents which indicated that the MSP meeting can be held during both the rainy and dry season.

CONCLUSIONS AND RECOMMENDATIONS

The study described multi-stakeholder platform participation based on key demographic indicators. There was high willingness of participation on the platform among smallholder rice farmers in Northern Ghana. The Probit regression analysis revealed that age of household head, household size and income were the main determinants of farmers' willingness to the MSP participation. Distance to meeting place was the most important constraint faced by the smallholder rice farmers in relation to their willingness to participate on the platform. Smallholder farmers residing at distant place from the meeting grounds were less likely to participate on the platform. The facilitators of the MSP must target and educate younger household heads within the region to ensure maximum participation and sustainability of the platform. Secondly, in designing of agricultural programmes, policy makers must incorporate activities that will impact positively on the income of farmers. Finally, the MSP must be situated at a central location coupled with improvements in road infrastructure in order to lessen the transportation burden of participants and consequently increase willingness of participation.

AUTHORS' CONTRIBUTIONS

This work was carried out in with the contribution from all listed authors. EM designed the study, performed the statistical analysis, and wrote the first draft of the manuscript. PME was in charge of the collection and management of the data. ANW managed and streamlined the analyses of the study and author WD managed the literature searches and edited the manuscript. All authors read and approved the final manuscript.

ACKNOWLEDGEMENTS

The authors greatly acknowledged financial support by USAID and the smallholder rice farmers that voluntarily participated in the survey.

REFERENCES

1. Adimado S (2001) Willingness to pay for research findings: a case study of pineapple farmers in Ghana. University of Ghana, Legon, Accra Ghana.
2. Asante BO, Afari-Sefa V, Sarpong DB (2011) Determinants of small-scale farmers' decision to join farmer based organizations in Ghana. Afr J Agric Res 6(10):2273-2279
3. Ayamga M (2006) Factors influencing the decision to participate in microcredit programmes: an illustration for Northern Ghana. Ghana J Dev Stud 3(2):57-65
4. Coase RH (1937) The nature of the firm. Economica 4:386-405
5. Coase RH (1960) The problem of social cost. J Law Econ 3:1-44
6. Delgado C (1999) Sources of growth in smallholder agriculture in sub-Saharan Africa: the role of vertical integration of smallholders with processors and marketers of high value-added items. Agrekon 38(S1):165-89
7. Dorward A, Kydd J, Poulton C, Bezemer D (2009) Coordination risk and cost impacts on economic development in poor rural areas'. J Dev Stud 45(1):1-20
8. Edwards AL (1964) Statistical methods for the behavioural sciences. Holt, Rmehart and Wintson, New York.
9. Enete AA, Igbokwe EM (2009) Cassava market participation decision of household in Africa. Tropicultura 27(3):129-136
10. Etwire PM, Dogbe W, Wiredu AN, Martey E, Etwire E, Owusu RK, Wahaga E (2013) Factors influencing farmer's participation in agricultural projects: the case of the agricultural value chain mentorship project in the Northern Region of Ghana. J Econ Sustain Dev 4(No.10):36-43
11. [http://www.statsghana.gov.gh/docfiles/glss5_report.pdf]

12. Ghana Statistical Service, GSS (2008) Ghana living standards survey report of the fifth round (GLSS 5), Accra, Ghana:,
13. Gockowski J, Ndoumbe M (2004) The adoption of intensive monocrop horticulture in Southern Cameroon. J Agric Econ 30:195-202
14. Goetz S (1992) A selectivity model of household food marketing behavior in sub-Saharan Africa. Am J Agr Econ 74:444-52
15. Greene HW (2003) Econometric analysis. Pearson Education, Inc., Upper Saddle River, New Jersey, USA.
16. Holloway G, Nicholson C, Delgado C, Staal S, Ehui S (2000) Agro-industrialization through institutional innovation, transaction costs, cooperatives and milk market development in the East African Highlands. Agr Econ 23(3):279-88
17. Key N, Sadoulet E, de Janvry A (2000) Transaction costs and agricultural household supply response. Am J Agr Econ 82(1):245-59
18. Kheralla M, Minet N, Kachule R, Souce BG, Berry P (2001) Impact of Agricultural Market Reforms on Smallholder Farmers in Benin and Malawi, Research Report, Vol. 2. IFPRI,
19. Langyintuo AS, Mekuria M (2005) Accounting for neighborhood influence in estimating factors determining the adoption of improved agricultural technologies. A paper at American Agricultural Economics Association annual meeting, providence. Rode Island,, pp 1–28
20. Makhura M, Kirsten J, Delgado C (2001) Transaction costs and smallholder participation in the maize market in the Northern Province of South Africa. Agriculture in Ghana: Facts and Figures, Accra, Ghana.
21. Markelova H, Meinzen-Dick R, Hellin J, Dohrn S (2009) Collective action for smallholder market access'. Food Pol 34:1-7
22. Martey E, Wiredu AN, Asante BO, Annin K, Dogbe W, Attoh C, Al-Hassan RM (2013) Factors influencing participation in rice development projects: the case of smallholder rice farmers in Northern Ghana. Int J Dev Econ Sustain 1(2):13-27
23. Mattson DE (1986) Statistics-Difficult Concept of Understanding Explanations'. Bolchanzy Carducci Publishers Inc,, p 281, 283.361,423

24. Ministry of Food and Agriculture, MoFA (2010) Agriculture in Ghana: facts and figures., Accra, Ghana, p 53
25. Mussei A, Mwanga J, Mwangi W, Verkuijl H, Mungi R, Elang A (2001) Adoption of improved wheat technologies by small-scale farmers in Mbeya District, Southern Highlands, Tanzania. International Maize and Wheat Improvement Centre (CIMMYT) and the United Republic of Tanzania, Mexico D.F.
26. Nnadi FN, Akwiwu CD (2008) "Determinants of Youths' Participation in Rural Agriculture in Imo State, Nigeria". Journal of Applied Sciences 8(2):328-333
27. Ofori IM (1973) Factors of agricultural growth in West Africa. ISSER, University of Ghana, Legon, Accra, Ghana.
28. Olwande J, Mathenge M (2010) Market participation among the poor rural households in Kenya. Tegemeo Institute of Agricultural Policy and Development,
29. Omamo SW (1998) Farm-to-market transaction costs and smallholder agriculture: Explorations with a non-separable household model. J Dev Stud 35:152-63
30. [www.ifpri.org/events/seminars/SmallFarms/] Pingali P, Meijer M, Khwaja Y (2005) Commercializing small farms: reducing transaction costs.:, Accessed 10 October 2007
31. Rölling N, Leeuwis C, Pyburn R (2002) Beyond the aggregation of individual preferences: moving from multiple cognition in resource dilemmas. In: Leeuwis C, Pyburn R (eds) Wheelbarrows full of frogs: social learning in rural resource management, Koninklijke Van Gorcum, Assen. pp 25-47
32. Schultz TW (1945) Agriculture in an unstable economy. McGraw-Hill Book Company Inc, New York.
33. Shepherd A (1997) Market information services: theory and practice. FAO (Food and Agriculture Organization), Rome.
34. Southworth HM, Johnston BF (1967) Agricultural development and economic growth. Cornell University Press, UK.
35. Stiglitz JE (1988) Economic organization, information and development. In: Chenery H, Srinivasan (eds) Handbook of development economics, vol. 1, chapter 5, Elsevier Science, Amsterdam.

36. Thiele G, Devaux A, Reinoso I, Pico H, Montesdeoca F, Pumisacho M, Andrade-Piedra J, Velasco C, Flores P, Esprella R, Kurt Manrique T, Horton D (2011) Multi-stakeholder platforms for linking small farmers to value chains: evidence from the Andes". Int J Agr Sustain:423–433
37. Thiele G, Oros R, Velasco C, Ambrose K (2005) Por qué trabajar con plataformas? In: Thiele G, Bernet T (eds) Conceptos, Pautas y Herramientas, CIP, Lima. pp 66-70171
38. Vellema S, Ton G, Nakimbugwe D, Mwesige D (2011) The Ugandan Oil Seed Sub-sector. Putting heads together: Agricultural innovation platforms in practice.
39. (2006) Where is the wealth of nations? Measuring capital for the 21st century. World Bank, Washington, DC.

Chapter 2

The Use of Adjuvants to Improve Uptake of Phosphorous Acid Applied to Pinus Radiata Needles for Control of Foliar Phytophthora Diseases

Carol Rolando[1], Robyn Gaskin[2], David Horgan[2], Nari Williams[1], and Martin K-F Bader[1]

[1]Scion, 49 Sala Street, Fenton Park, Private Bag 3020, Rotorua 3046, New Zealand

[2]Plant Protection Chemistry NZ, 49 Sala Street, Fenton Park, Rotorua 3043, New Zealand

ABSTRACT

Background

Phosphorous acid is being investigated as a fungicide for the management of a needle disease caused by *Phytophthora pluvialis* in *Pinus radiata* in New Zealand. However, little is known about the penetration characteristics of this fungicide into *Pinus radiata* foliage. This study was undertaken to determine: i) the penetration characteristics of a commercial phosphorous acid formulation, applied at 3 kg ha^{-1} and 12 kg ha^{-1} in 100 L water, into *Pinus radiata* foliage and, ii) the effect of four commercially available adjuvants on phosphorous acid uptake into *Pinus radiata*foliage. Efficacy of the best treatment was tested in vitro with two *Phytophthora* species,*Phytophthora kernoviae* and *P. pluvialis*.

Methods

Foliar uptake of orthophosphoric acid radio-labelled with [^{32}P] was used to determine penetration characteristics of phosphorous acid as affected by the adjuvants tested. Needles collected from potted *Pinus radiata* plants sprayed with the best performing treatment were used in a detached needle assay, where treated needles were exposed to the zoospores of the *Phytophthora* species being tested.

Results

Uptake of the phosphorous acid formulation into *Pinus radiata* needles applied without adjuvants was low (3.9-6.6%) at both concentrations tested. An alcohol ethoxylate adjuvant (applied at 0.2%) and two organosilicone adjuvants (applied at 0.2%) were found to significantly increase uptake of the fungicide over that applied alone or in combination with pinolene (applied at 0.35%). Highest uptake of phosphorous acid into *Pinus radiata* foliage (51.6%) occurred over a period of 72 hours when the phosphorous acid was applied at the equivalent of 12 kg ha^{-1} with an organosilicone blend adjuvant at 0.2%. A significant reduction in lesion length on infected needles was found relative to the control when the two phosphorous acid treatments were applied with 0.2% organosilicone blend adjuvant. However, there were no significant differences in lesion length between the two treatments.

Conclusions

These results show that an organosilicone adjuvant significantly enhances uptake of phosphorous acid into *Pinus radiata* needles and may increase its efficacy against *Phytophthora* species.

BACKGROUND

A new needle disease was reported in *Pinus radiata* D. Don stands on the east coast of New Zealand during routine forest surveillance in late April 2008. Observations of the symptoms, their development and the apparent spread of the disease, known locally as red needle cast, all gave a strong indication of a likely biotic cause, which led to further investigations. Isolations made from symptomatic needles yielded an undescribed species of *Phytophthora* as the likely causal pathogen (Dick et al. [2014]). The causal pathogen has recently been described as *Phytophthora pluvialis*which is considered to be endemic to Oregon, USA (Reeser et al. [2013]). Subsequently, a second species *Phytophthora kernoviae* Brasier, Beales & Kirk, was also recovered from needles, though at a consistently lower rate (Dick et al. [2014]). Concurrent to epidemiological work for red needle cast, a range of trials were initiated to investigate possible fungicides that could be effective for management of red needle cast on *Pinus radiata* in New Zealand. Phosphorous acid was chosen as a key fungicide for further research on potential red needle cast management since this active ingredient (in the form of phosphorous acid) is used widely in agriculture and horticulture for control of *Phytophthora* species (Lobato et al. [2011]; Rebollar-Alviter et al. [2007]; Shearer and Crane [2009]).

Phosphorous acid fungicides, applied either pre-infection or for protectant activity, are the only known ambi-mobile fungicides capable of moving upward and downward in plants (Cohen and Coffey [1986]). Phosphorous acid exhibits a complex mode of action, acting both directly on the pathogen and also indirectly by stimulating host defence responses to ultimately inhibit pathogen growth (Tomlin

[2006]). The fungicide can persist in host plants for several months post application, however persistence is host dependent (Tynan et al. [2001]). When treated, most plants exhibit phytotoxic effects above a (species dependent) threshold concentration, with field treated plants more sensitive to phytotoxic effects than those treated in a glasshouse (Hardy et al.[2001]). Kanaskie et al. ([2011]) showed the potential for phosphorous acid to control outbreaks of sudden oak death, caused by *Phytophthora ramorum* S. Werres, A.W.A.M. de Cock & W.A. Man, in*Notholithocarpus densiflorus* (Hook. & Arn.) Manos, Cannon & S.H. Oh (south-western Oregon tanoak). Besides the work described by Kanaskie et al. ([2011]), there are few reports describing the use of phosphorous acid to control *Phytophthora* spp. associated foliar diseases in mature trees. This is likely because there are few tree-infecting aerial *Phytophthora* spp. currently known and those which are known (such as *P. ramorum, P. kernoviae, P. pinifolia* and *P. pluvialis*) have been relatively recently observed. Graham ([2011]) suggested options for using phosphorous acid to control *Phytophthora* species in forest trees based on work done on citrus orchards. Phosphorous acid has been successfully used to control *Phytophthora* root rot in *Pinus radiata*nurseries (Reglinski et al. [2009]). However, the results of this study cannot readily be extrapolated to airborne *Phytophthora* pathogens infecting mature trees.

Maximising uptake of pesticides is important to optimise efficacy and cost-effectiveness of any chemical control strategy. However, little is known about the penetration characteristics of phosphorous acid into *Pinus radiata* foliage. Agrochemical adjuvants are commonly used to improve efficacy of systemic pesticides by improving spray deposition and retention on the target species and/or by increasing the penetration of the pesticides into the plant (Zabkiewicz [2000]).

This study was undertaken to determine: i) the penetration characteristics of a commercial phosphorous acid formulation into *Pinus radiata* foliage and, ii) the effect of commercially available adjuvants on phosphorous acid uptake into *Pinus radiata* foliage. Efficacy of the best treatment was subsequently tested in an *in-vitro* bioassay conducted with the pathogens, *Phytophthora kernoviae* and *P. pluvialis*. Selection was based on a combination of adjuvant chemistry, their mode of action, and current relative cost. Four adjuvants, representing three classes of adjuvant, were selected for testing: i) a low-cost pinolene that is used as a wetter, sticker and rainproofer, ii) a moderately priced non-ionic surfactant typically used to aid wetting, sticking and penetration of fungicides, and iii) two relatively expensive organosilicones that are known to improve retention and coverage of sprays, and are often utilised to increase uptake of systemic pesticides into plants.

METHODS

Three clones (referred to as A, B and C) of grafted *Pinus radiata,* hereafter referred to as the potted plants, known to be susceptible to the pathogens *Phytophthora kernoviae* and *P. pluvialis,*were used in this investigation. The potted plants, supplied from the Radiata Pine Breeding Company Ltd. (Rotorua, New Zealand), were obtained in mid-autumn from the Scion nursery (Rotorua, New Zealand) and were equilibrated for two weeks in a controlled environment cabinet (20/15°C, 70% relative humidity, 12 h photoperiod, 500 µmol m^{-2} sec^{-1} light) prior to treatment application. All treatments were applied to one-year-old needles located on the main leader of the potted plants. There were three potted plants of each clone for each treatment, such that all treatments were replicated three times, with three experimental units per clone (three clones x three potted plants of each clone).

A preliminary experiment was undertaken to gauge the uptake of phosphorous acid into *Pinus radiata* needles without the use of an adjuvant. As there is no label rate for aerial application of phosphorous acid to mature *Pinus radiata* plantations in New Zealand, the rates used were obtained from aerial application rates used in Western Australia, for *Lambertia inermis* var. *inermis*and *Banksia grandis,* and North America for *N. densiflorus* (Kanaskie et al. [2011]; Shearer and Crane [2009]). Phosphorous acid (Agrifos®600, Key Industries; 600 g L^{-1} phosphorous acid) was applied in 100 L ha^{-1} of water at two rates (3 kg ha^{-1} and 12 kg ha^{-1}), Treatments 1 & 2, Table 1. Based on the results of this initial test, it was decided to use the highest application rate of phosphorous acid in further experiments to increase the likelihood of active ingredient being absorbed by the foliage.

Table 1: Uptake of [32P] phosphorous acid into *Pinus radiate* needles with or without various adjuvants at 24 hours after treatment (HAT) and droplet spread on needle surface

Treatment no	Rate of active (kg ha-1)	Adjuvant	Adjuvant concentration	[32P] uptake (%)	[32P] uptake dose* (kg ha-1)	Spread area of droplets (mm2)
1	3	None	-	6.6 cd	0.20	0.8 h
2	12	None	-	3.9 d	0.47	1.0 g
3	12	Pinolene	0.35	6.8 cd	0.82	1.2 f
4	12	Alcohol ethoxylate	0.05	7.7 cd	0.92	1.5 e
5	12	Alcohol ethoxylate	0.10	18.6 b	2.23	1.7 d
6	12	Alcohol ethoxylate	0.20	14.2 bc	1.70	2.0 c
7	12	Organosilicone penetrant	0.20	20.1 b	2.41	2.3 b
8	12	Organosilicone blend	0.20	36.9 a	4.43	2.7 a

Means within columns sharing common postscripts are not significantly different.

*Nominal dose calculated as (phosphorous acid rate x %_uptake).

Rolando *et al.*

Rolando *et al. New Zealand Journal of Forestry Science* 2014 44:8 doi:10.1186/s40490-014-0008-5

Experiment 1: Effect of Four Adjuvants on Uptake of Phosphorous Acid

Description of Adjuvants

The adjuvants tested were i) a pinolene (Nu-Film-17®, Key Industries Ltd, Auckland), ii) a linear alcohol ethoxylate (950 g L^{-1}, Actiwett®, Etec Crop Solutions Ltd, Auckland), iii) a pH-stable organosilicone polymer with super-penetrant activity (Gaskin et al. [2012]) containing alcohol ethoxylate and polyalkylene compounds (Du-Wett® Stainless, Etec Crop Solutions Ltd, Auckland) and iv) an organosilicone-blend containing siloxane polyalkeneoxide copolymers (Du-Wett®, Etec Crop

Solutions Ltd, Auckland) designed for use as an horticultural super-spreader (AgriMedia Ltd[2013]). All adjuvants used in the trial were tested within manufacturer's recommended rates or at rates known to be effective for penetration into horticultural and agricultural crops.

Determination of Treatment Penetration

The penetration characteristics of phosphorous acid were tested for up to 24 h after treatment (HAT). Phosphorous acid was applied as Agrifos® 600, at the equivalent of 12 kg ha^{-1} in 100 L ha^{-1} water in combination with the four commercially available adjuvants described in 2.1.1. Details of the adjuvant concentrations (Treatments 3–8) are given in Table 1.

Orthophosphoric acid radio-labelled with [^{32}P] (ARC, Inc., USA) was added to freshly prepared treatment solutions in deionised water 30 minutes prior to use to enable the radiaoactive chemical to equilibrate with the mono-and di-potassium salts present in the commercial phosphorous acid formulation. The [^{32}P] phosphoric acid comprised <1% mass of the total phosphorous acid in solution at all times. Droplets (10 × 0.24 µL) of the radiolabelled treatment solution were applied using a microsyringe to the central region of a single needle per fascicle such that they simulated a 100 L ha^{-1} spray application volume (i.e. 1 µL cm^{-2}). Treated fascicles were pre-tagged with coloured electrical wire for identification.

The quantity of radiolabelled phosphorous acid applied to each needle in each treatment was determined by dispensing droplets, as applied to plants, directly into scintillation vials (3 replicates). Following treatment application, needle fascicles were harvested at 24 HAT (Table 1). The entire fascicle, including the treated leaf surface, was washed with a mixture of water and ethanol (1:1 v/v; 2 × 4 mL) to recover any unabsorbed phosphorous acid. Prior audits determined that such wash-offs could recover >95% of applied phosphorous acid solutions from foliage surfaces at dry-down. Foliage in Treatment #3 (Table 1) was given an additional wash with 50% aqueous acetone to ensure all phosphorous acid remaining on the surface was recovered. This extra wash was carried out because of the low solubility of the pinolene adjuvant in aqueous solution.

Scintillant solution (13 mL ACS; Amersham Biosciences, Australia) was added to the washings and radioactivity was quantified by liquid scintillation counting (Packard Tricarb 2100 TR). Foliar uptake was defined as the radioactivity not recovered from washing the treated needles and was calculated as a percentage of the applied dose.

Determination of Droplet Spread

In a separate test, all treatment solutions (1–8; Table 1) were prepared as for the penetration experiment but with the addition of a water-soluble fluorescent dye, (Blankophore P, Bayer) at a concentration of 0.5% w/v immediately prior to droplet application. Twelve droplets (0.24 μL) of each solution were applied to each of three needles on each potted plant (and different clones). Droplets were left to spread for 10 minutes. Needles were then illuminated with ultra-violet light and photographed. The area of droplet spread in each case was determined using image analysis software (V++™, Digital Optics Ltd).

Experiment 2: Optimising Uptake of Phosphorous Acid

The most promising adjuvant identified in the first experiment was the organosilicone-blend superspreader (Du-Wett®). The second experiment examined the effect of a wider range of application rates of this adjuvant on the uptake of phosphorous acid over a period of 24 to 72 h.

Phosphorous acid containing [^{32}P] was applied at the equivalent rate 12 kg ha^{-1} in 100 L water in combination with various rates of the organosilicone (Table 2). Treatments were applied to the one-year-old needles of the same three clones (A, B and C) of potted *Pinus radiata* used in Experiment 1 and two treatments from that experiment were repeated. Treatment application and determination of phosphorous acid uptake were as described for Experiment 1.

Table 2: [32P] phosphorous acid uptake into *Pinus radiata*using various rates of the organosilicone-blend superspreader adjuvant (Du-Wett®) at two harvest times

Treatment no	Adjuvant concentration (%)	Harvest time HAT	[32P] uptake (%)	[32P] uptake dose* (kg ha-1)
2R#	none	24	3.5 d	0.42
8R#	0.20	24	43.6 abc	5.23
9	0.05	24	31.8 c	3.82
10	0.10	24	33.6 bc	4.03
11	0.10	72	47.3 ab	5.68
12	0.20	72	51.6 a	6.19

Phosphorous acid was applied at the equivalent of 12 kg ha^{-1} in all treatments. Means sharing common postscripts are not significantly different.

#The postscript R indicates a repeat treatment.

*Nominal dose calculated as (phosphorous acid rate x %_uptake).

Rolando *et al.*

Rolando *et al.* *New Zealand Journal of Forestry Science* 2014 44:8 doi:10.1186/s40490-014-0008-5

Experiment 3: Testing the Efficacy of the Best Acid/Adjuvant

Combination against *Phytophthora* Species Using a Detached Needle Bioassay

Both rates of phosphorous acid tested previously (3 kg ha^{-1} and 12 kg ha^{-1}) were evaluated with the best-performing adjuvant Du Wett® added at 0.2% on the same three *Pinus radiata* clones (A, B and C). Treatments were applied in the equivalent of 100 L ha^{-1} using a calibrated, moving head tracksprayer (Plant Protection Chemistry NZ, Rotorua) fitted with twinjet nozzles (TJ60-8002EVS). There were a total of 27 potted plants in the trial; three for each of the two treatments (plus a control) with each clone. After treatment, the potted plants were sheltered from rain for 72 h after which they were placed outside and watered regularly.

Zoospore inoculum was prepared by growing isolates on carrot agar at 17°C for three days. Plugs of agar and mycelium were taken from the leading edge of the colonies, flooded with clarified carrot broth (Erwin & Ribeiro [1996]) and incubated for three days at 17°C. The resulting mycelial mats were rinsed thoroughly three times with sterile deionised water, drained and flooded with sterile pond water. Pond water was collected from a local pond and used in the development of*Phytophthora pluvialis* and *P. kernoviae* cultures. These were incubated in the dark for a further three days before zoospore release was induced with 45 minute intervals at 4°C in the dark at room temperature (21-22°C) on a light box. Zoospore concentrations were determined using a haemocytometer and standardised to 1×10^5 zoospores per mL. Zoospore suspensions were used within two hours of preparation.

Five days after treatment, 30 fascicles were collected from each potted plant; including the controls. Ten fascicles each, from each potted plant. were exposed to either a suspension in pond water of 1×10^5*P. pluvialis* or *P. kernoviae* zoospores per ml for 24 h. Controls were exposed to autoclaved pond water only. Following exposure, needles were placed on trays moistened with wet paper towels and incubated in a controlled environment (17°C, 65-70% relative humidity, 14 h photoperiod) for 10 days. After this time, the needles within each fascicle were separated and lesion length was measured.

Statistical Analyses

Treatments in Experiments 1 and 2 were compared by analysis of variance and least significant difference (LSD) using the statistical program Statistix 9 (Analytical Software, USA). Results from Experiment 3 (the detached needle bioassay) were analysed using the statistical programme R version 2.15.2 (R Development Core Team, 2012). A linear mixed effects model using maximum restricted likelihood estimation (R-package *nlme*) was applied to test for differences in the length of needle lesions between a zero control and phosphorous acid treatments. The fixed terms of the model contained 'phosphorous acid treatment' (factor with 3 levels: 0, 3 and 12 kg ha^{-1}) and '*Phytophthora* pathogen' (factor with 3 levels: control, *P. pluvialis* and *P. kernoviae*) and their interaction. 'Host tree identity', 'subsample identity' within tree and 'fascicle identity' within subsample were modelled as nested random effects. Tukey contrasts (R-package *multcomp*) were used to compare the effects of the phosphorous acid treatment for each *Phytophthora* species separately.

RESULTS AND DISCUSSION

Experiment 1: Effect of Four Adjuvants on Uptake of Phosphorous Acid

Comparison of Adjuvants

The results of the preliminary experiment indicated that uptake of phosphorous acid into *Pinus radiata* needles was low and proceeded very slowly. A similar uptake of phosphorous acid was recorded at 24 h after treatment for the 3 kg ha^{-1} and 12 kg ha^{-1} phosphorous acid application rates, 6.6% and 3.9% respectively (Treatments 1 & 2; Table 1). Based on this preliminary experiment, it was decided to proceed with the highest rate of phosphorous acid tested in the experiment (treatments 3–8), to maximise the acid dose absorbed by the foliage.

Addition of 0.35% pinolene as an adjuvant did not increase uptake of phosphorous acid (Treatment 3; Table 1). This was not unexpected since this type of polymer is claimed to provide increased resistance of sprays to rain wash-off and improved adhesion on difficult-to-wet plants, but is not known to increase pesticide uptake.

Addition of 0.05% alcohol ethoxylate (Treatment 4) was ineffective but the addition of the same adjuvant at 0.1% (Treatment 5) significantly increased uptake of phosphorous acid. There was no additional benefit from using a higher concentration of 0.2% (Treatment 6; Table 1). This class of adjuvant is known to improve the adhesion and spreading properties of sprays, and has been shown to promote increased uptake of a range of compounds on a variety of species (Stock et al.[1992]). These authors also showed that the magnitude of increased penetration varied according to properties of both the compound and the adjuvant and was influenced by surfactant concentration and target plant species.

The two organosilicone adjuvants (Du-Wett® and Du-Wett® Stainless) were used at the same rate (0.2%) (Treatments 7 & 8) and both significantly increased uptake relative to the other adjuvants tested. The organosilicone blend (superspreader adjuvant (Du-Wett®); Treatment 8) was the most effective of all the treatments tested, increasing uptake to nine fold greater than that of the phosphorous acid used alone (Table 1). This organosilicone blend was used at a higher rate of addition than when typically used as a superspreader in horticultural operations (≤ 0.1%) in order to maximise its penetration properties.

The pH-stable organosilicone penetrant (Du Wett® Stainless) was included in the study because it is recommended for use with acidic active ingredients (AgriMedia Ltd [2013]; Policello and Gaskin[2010]) such as phosphorous acid. However, the formulated phosphorous acid used in this study, Agrifos® 600, is buffered to moderately neutral (pH of 6.8) meaning use of a pH-stable organosilicone penetrant is not essential. This was useful since the pH-stable organosilicone penetrant was more expensive than the organosilicone blend.

Standard organosilicone superspreader adjuvants should only be used at neutral pH (6–8) as these adjuvants degrade rapidly in acid/alkaline environments (Policello et al. [1995]). The superspreader adjuvant used in the current trial is disclosed as a proprietary blend of organosilicone and organic fluids and is not typically used as a penetrant. However, it contains polymers with uptake-enhancing capability when combined, at appropriate concentrations, with phosphorous acid.

Droplet Spread

The spread of droplets on a leaf surface is a measure of how well a spray can be expected to cover the plant surface following droplet adhesion. *Pinus radiata* foliage is classified as easy-to-wet, with a surface contact angle of 64° (Gaskin et al. [2005]). Surfactants are most useful for improving droplet spread and adhesion on difficult-to-wet species (Gaskin et al. [2005]). The effect of adjuvants on droplet spread on easy-to-wet foliage confirmed this (Table 1). Although significant, pinolene increased the spread of phosphorous acid spray droplets negligibly. Alcohol ethoxylate improved droplet spread and the effect increased with increasing concentration (Table 1). The pH-stable organosilicone provided slightly enhanced droplet spreading relative to either pinolene or alcohol ethoxylate. The most effective treatment, however, was the organosilicone blend superspreader, which significantly increased coverage of needle surfaces by almost three-fold compared to the controls.

Experiment 2: Optimising Uptake of Phosphorous Acid

A repeat of the control phosphorous acid treatment at a rate of 12 kg ha^{-1} confirmed that very little penetration of phosphorous acid occurred in the absence of adjuvant (Treatment 2R; Table 2). A repeat of Treatment 8 from Experiment 1 (8R, Table 2) confirmed that the organosilicone blend increased phosphorous acid uptake. This treatment and all the others using the organosilicone blend superspreader (Treatments 9–12) increased uptake by more than nine-fold relative to the control (Table 2). The highest uptake occurred at the highest concentration tested (0.2%) at both 24 h and 72 h after treatment.

Uptake with the lowest concentration of organosilicone blend (0.05%) showed high variability within replicates and, as a result, this treatment was repeated with similar variability in results. The reasons for this are not obvious, but suggest that a low concentration of organosilicone blend adjuvant will provide highly variable uptake of phosphorous acid into *Pinus radiata* foliage.

An analysis of treatments 8R, 10, 11 & 12 (Table 2) as a factorial of adjuvant concentration (0.1% and 0.2%) and harvest time (24 h and 72 h) showed that the main effects of both concentration and uptake over time were significant ($p<0.05$). Thus, uptake of phosphorous acid is expected to be greater with the higher concentration of organ silicone blend and to proceed for up to 72 HAT at least (Table 2). Droplets of all treatments were still visibly hydrated on foliage at 24 HAT, whereas at 72 HAT all droplets had visibly dried. Uptake is likely to cease when fungicide deposits become dry (Merritt [1982]) so the rate of uptake will slow with time.

Testing the Efficacy of Best Acid/Adjuvant Combination Against *Phytophthora* Species

Lesion length of *Pinus radiata* needles inoculated with either *Phytophthora kernoviae* or *P. pluvialis*was significantly reduced by application of the phosphorous acid treatments containing 0.2% organosilicone blend adjuvant at both application rates tested compared to controls (Figure 1). However, there was no significant difference in lesion length between the low- and high-dose treatments used and either *Phytophthora* species tested. This result indicated that sufficient phosphorous acid had been absorbed by the needles to affect pathogen activity at both concentrations. In addition, these results indicate that phosphorous acid has activity against the target *Phytophthora* species when it is applied with organosilicone blend adjuvant. Similar bioassays have been used extensively to demonstrate the efficacy of phosphorous acid application to imply field based control for a range of *Phytophthora* diseases (Barrett et al. [2003]; Garbelotto and Schmidt [2009]; Shearer et al. [2012]). While bioassays can provide a useful indication of treatment efficacy, in some cases, they have also been shown to differ significantly from field based analyses and are therefore no substitute for field based results (Shearer et al. [2012]). Therefore, field trials involving application of phosphorous acid to mature *Pinus radiata* trees need to be carried out to determine an optimum dose for management of needle diseases in the field.

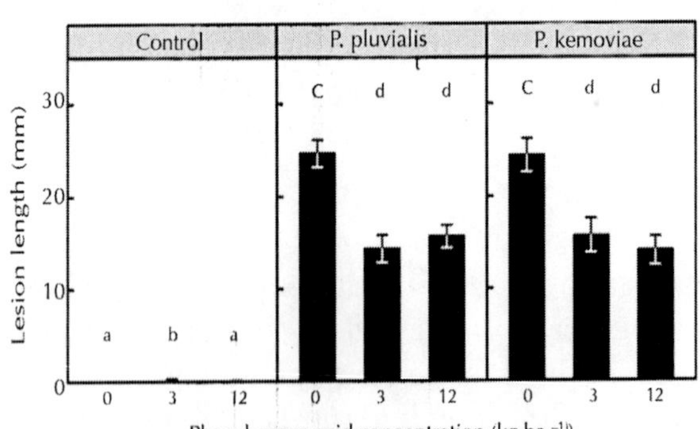

Figure 1: Lesion length (model predictions and confidence intervals) on*Pinus radiata*needles exposed to*Phytophthora pluvialis*or*Phytophthora kernoviae*-

for 24 hours following treatment with either 3 kg ha^{-1}or 12 kg ha^{-1}phosphorous acid each applied in 100 L water containing 0.2%organosilicone-blend superspreader adjuvant (Du-Wett®).

Kanaskie et al. ([2011]) conducted field trials to test the efficacy of phosphorous acid for the control of sudden oak death in south western Oregon tanoak. In aerial spray trials phosphorous acid was applied at two rates, 17.36 kg ha^{-1} and 34.72 kg ha^{-1}, in conjunction with Pentra-bark® in 160 L water ha^{-1}. Detached-twig and log bioassays with *P. ramorum*, as well as *in situ* stem inoculations with *Phytophthora gonapodyides* (Petersen) Buisman, were carried out to investigate treatment efficacy. Aerial spray with phosphorous acid at both application rates consistently resulted in smaller bole lesions on trees challenge inoculated with *P. gonapodyides in situ* and logs inoculated with *P. ramorum*. However, results from detached canopy twig assays were variable and showed only small treatment effects. These findings demonstrate the potential variability of field based treatments and the need to assess treatment effects both in-field and using bio-assays tested across different tissues. In the present study, bioassays focussed on detached needle inoculations as red needle cast has, as yet, only been shown to affect *Pinus radiata* needles (Dick et al. [2013]). Further confirmation of these results in the field is essential to establish the efficacy of broad-scale treatment for the control of red needle cast in *Pinus radiata*.

CONCLUSIONS

There was very little penetration of phosphorous acid into the foliage of *Pinus radiata* 24 hours after treatment in the absence of adjuvants. The addition of either an alcohol ethoxylate or organosilicone adjuvant improved uptake. The highest uptake of phosphorous acid occurred with an organosilicone blend adjuvant (Du-Wett®) at a concentration of 0.2%, and uptake progressed over a period of more than 24 hours after treatment. The addition of either of these adjuvants improved surface coverage of phosphorous acid sprays and they are also likely to have some beneficial effects on spray retention.

It is clear that the use of an organosilicone blend adjuvant in phosphorous acid treatments will improve coverage of sprays on *Pinus radiata*, in addition to its effects on the penetration of phosphorous acid into the plants. Its use is recommended to increase the uptake of phosphorous acid when applied to *Pinus radiata* for management of red needle-cast disease. Preliminary indications from the detached needle bioassays indicate that the rates of phosphorous acid used in this investigation (3 kg ha^{-1} and 12 kg ha^{-1}) have the potential to reduce infection of *Pinus radiata*by *Phytophthora* species including *P. kernoviae* and *P. pluvialis*. Field trials are needed to confirm this.

AUTHORS' CONTRIBUTIONS

CR and RG lead the conceptual development of this paper. RG and DH carried out all experimental work to characterise the penetration of phosphite as affected by four commercially available adjuvants. CR designed the study to test the best performing treatment on potted *P. radiata*, with NW the conceptual lead on work associated with in-vitro testing of *P. kernoviae* and *P. pluvilais*. MB analysed data and developed models to present Figure 1. All authors read and approved the final manuscript.

ACKNOWLEDGEMENTS

The Forest Owners Association, New Zealand, and Ministry of Business Innovation and Employment, New Zealand, provided the funding for this work.

REFERENCES

1. (2013). New Zealand Novachem Agrichemical Manual. Ashcroft House, Christchurch, New Zealand.
2. Barrett, S, Shearer, B, Hardy, GESJ (2003). The efficacy of phosphite applied after inoculation on the colonisation of *Banksia brownii* sterns by *Phytophthora cinnamomi*. *Australasian Plant Pathology, 32*, 1–7.
3. Cohen, Y, & Coffey, MD (1986). Systemic fungicides and the control of oomycetes. *Annual Review Phytopathology, 24*, 311–388.

4. Dick, M, Williams, NM, Hood, IA, Bader, M. (2014). Pathogenicity of Phytophthora pluvialis to Pinus radiata and its relation with red needle cast disease in New Zealand. New Zealand Journal of Forestry Science.44, 6.
5. Erwin, DC, & Ribeiro, O (1996). Phytophthora Diseases Worldwide. APS Press, Minnesota.
6. Garbelotto, M, & Schmidt, DJ (2009). Phosphonate controls sudden oak death pathogen for up to two years. *California Agriculture*, *63*, 10–17.
7. Gaskin, RE, Steele, KD, Forster, WD (2005). Characterising plant surfaces for spray adhesion and retention. *New Zealand Plant Protection*, *58*, 197–183.
8. Gaskin, RE, Logan, DP, May, W, Rowe, CA, Steele, KD, van Leeuwen, RM, Connolly, PG (2012). Control of passionvine hopper and cicada eggs on kiwifruit canes with bifenthrin and a new super-penetrant adjuvant. *New Zealand Plant Protection*, *65*, 100–105.
9. Graham, JH (2011). Phosphite for control of Phytophthora diseases in citrus: model for management of *Phytophthora* species on forest trees? *New Zealand Journal Forestry Science*, *41S*, S49–S56.
10. Hardy, GESJ, Barrett, S, Shearer, BL (2001). The future of phosphite as a fungicide to control the soilborne plant pathogen *Phytophthora cinnamomi* in natural ecosystems. *Australasian Plant Pathology*, *30*, 133–139.
11. Kanaskie, A, Hansen, EM, Sutton, W, Reeser, P, Choquette, C (2011). Application of phosphonate to prevent sudden oak death in south-western Oregon tanoak (*Notholithocarpus densiflorus*) forests. *New Zealand Journal of Forestry Science*, *41S*, S177–S187.
12. Lobato, MC, Machinandiarena, MF, Tambascio, C, Dosio, GAA, Caldiz, DO, Daleo, GR, Andreu, AB, Olivieri, FP (2011). Effect of foliar applications of phosphite on post-harvest potato tubers. *European Journal of Plant Pathology*, *130*, 155–163.
13. Merritt, CR (1982). The influence of form of deposit on the phytotoxicty of MCPA, paraquat and glyphosate applied as individual drops. *Annals of Applied Biology*, *101*, 527–532.

14. Policello, GA, & Gaskin, RE In Baur P, Bonnet M (Eds.) (2010). Hydrolysis resistant organosilicon adjuvant for glyphosate. *Proceedings of the 9th International Symposium on Adjuvants for Agrochemicals Techinical University of Munich* Freising, Germany.

15. Policello, GA, Stevens, PJG, Forster, WA, Murphy, GJ In Hall FR, Berger PD, Collins HM (Eds.) (1995). The influence of pH on the performance of organosilicone surfactants. *Pesticide Formulations and Application Systems: 14th Volume* (pp. 313–317). ASTM, West Conshohocken, PA, USA.

16. Rebollar-Alviter, A, Madden, LV, Ellis, MA (2007). Pre- and post-infection activity of azoxystrobin, pyraclostrobin, mefenoxam, and phosphite against leather rot of strawberry, caused by*Phytophthora cactorum*. *Plant Disease, 91*, 559–564.

17. Reeser, PW, Sutton, W, Hansen, EM. (2013). Phytophthora pluvialis, a new species found in mixed tanoak – Douglas-fir forests of western Oregon, U.S.A. North American Fungi 8(7), 1–8.

18. Reglinski, T, Spiers, TM, Dick, MA, Taylor, JT, Gardner, J (2009). Management of phytophthora root rot in radiata pine seedlings. *Plant Pathology, 58*, 723–730.

19. Shearer, BL, & Crane, CE (2009). Influence of site and rate of low-volume aerial phosphite spray on lesion development of *Phytophthora cinnamomi* and phosphite persistence in *Lambertia inermis*var. *inermis* and *Banksia grandis*. *Australasian Plant Pathology, 38*, 288–304.

20. Shearer, BL, Crane, CE, Scott, PM, Hardy, GES (2012). Variation between plant species of in-planta concentration and effectiveness of low-volume phosphite spray on *Phytophthora cinnamomi*lesion development. *Australasian Plant Pathology, 41*, 505–517.

21. Stock, D, Edgerton, BM, Gaskin, RE, Holloway, PJ (1992). Surfactant-enhanced foliar uptake of some organic compounds: Interactions with two model polyoxyethylene aliphatic alcohols. *Pesticide Science, 34*, 233–242.

22. Tomlin, C (2006). The Pesticide Manual. British Crop Protection Council, Hampshire.

23. Tynan, KM, Wilkinson, CJ, Holmes, JM, Dell, B, Colquhoun, IJ, McComb, JA, Hardy, GEJ (2001). The long-term ability of phosphite to control *Phytophthora cinnamomi* in two native plant communities of Western Australia. *Australian Journal of Botany, 49*, 761–770.

24. Zabkiewicz, JA (2000). Adjuvants and herbicidal efficacy - Present status and future prospects.*Weed Research*, *40*, 139–149.

Chapter 3

Assessing Hydrological And Provisioning Ecosystem Services in a Case Study in Western Central Brazil

Lars Koschke[1], Carsten Lorz[2], Christine Fürst[3], Tobias Lehmann[1], and Franz Makeschin[1]

[1]Department of Soil Science and Site Ecology, Dresden University of Technology, Pienner Str. 19, 01737 Tharandt, Germany

[2]Faculty for Woods and Forestry, Forest Pedology, Geology and Site Ecology, HochschuleWeihenstephan-Triesdorf, University of Applied Sciences, Freising, Germany

[3]Center for Development Research, Department of Ecology and Natural Resources Management, University of Bonn, Walter-Flex-Str. 3, 53113 Bonn, Germany

ABSTRACT

Introduction

Current land use in the Federal District, Western Central Brazil, causes problems related to the water supply which are linked to the regulation of ecosystem services (ES). In scope of an Integrated Water Resources Management concept, we further developed the web-based planning support tool GISCAME for the Pipiripau river basin case study.

Methods

We introduced analyses on ecosystem potentials in the raster-based tool to assess, in a spatially explicit manner, the scenario impact on water purification, sediment retention, water retention, and provision of food and fodder in order to identify potential pathways for conserving water resources. To demonstrate the method, we assess ES depending on a number of land use/land cover change (LULCC) scenarios.

Results

We found that a considerable increase of water purification and sediment retention is difficult to achieve with realistic small scale LULCC, mainly because in areas with a low potential to provide hydrological ES and thus with a high demand for sustainable land use, such as native Savanna (Cerrado) and natural forests (Mata), favorable land uses were often already existing. We observed synergies in the response of regulating hydrological ES to LULCC but at the same time also trade-offs with provision of food and fodder.

Conclusions

Our findings suggest that further degradation and loss of Cerrado and Mata must be avoided and their restoration should be promoted in order to safeguard water resources. We suggest that restoration measures should be focused on arable land located at steep slopes near surface waters to effectively increase hydrological ES through the marginal reduction of provision services.

INTRODUCTION

Urban sprawl and intensive agriculture in Western Central Brazil have led to large scale degradation and loss of Cerrado, the native Savanna vegetation (Schmidt et al. 2009). Current land use causes high loads of suspended solids in stream water and subsequent silting of drinking water reservoirs. These processes, leading to the deterioration of (raw) water quality, the loss of some smaller reservoirs, and substantial reductions of the volume of larger reservoirs, threaten the regional water resources and especially the provision of drinking water (Felizola et al. 2001; Fortes et al. 2007; Lorz et al. 2012a). The water supply company (CAESB) of the Distrito Federal (DF) noticed increasing water treatment costs in the meso-scale Pipiripau river basin in the north-eastern part of the DF due to soil erosion and nutrient runoff from adjacent agricultural areas (Buric and Gault 2011). Adapted land use is a means to reduce diffuse pollution of water resources and sediment generation as a result of agricultural practices (Schwab et al. 1995). In this sense, the river basin is part of the Water Producer Program (Programa Produtor de Água) that has been initiated by the National Water Agency of Brazil (ANA); the program aims to implement best management practices. Strategies mentioned in this program include, for instance, the restoration and conservation of natural vegetation in priority areas such as riparian zones (Strauch et al. 2013).

From the benefits that humans gain from ecosystems (ecosystem services, ES), many are related directly or indirectly to freshwater. Those ES that are closely related to the hydrological cycle are commonly referred to as hydrological or water related ES and comprise the provision of services such as the supply of water for drinking and irrigation, energy production, and regulating services such as water purification and erosion control (Brauman et al. 2007; de Groot et al. 2010). As the provision of drinking water mostly relies on the presence and status of native vegetation within a watershed (Pattanayak and Wendland 2007), the preservation and restoration of natural vegetation is supposed to be advantageous also for other ES and for biodiversity conservation (Pert et al. 2010). Provision of regulating hydrological ES is also governed by terrestrial characteristics of the watershed, such as slope, topography, and soil depth (Terrado et al. 2013). Further, the location of land use and land cover change (LULCC) is highly relevant for the performance of certain ES (Bryan and Crossman2008; Rounsevell et al. 2012).

We aim to support river basin management and focus on land use planning and sediment management in the Pipiripau river basin. The research questions presented are:

- Can generally available data on soil and topographic conditions provide a meaningful foundation for ES-based Integrated Water Resources Management (IWRM) and spatial planning support at the landscape level?
- Does the inclusion of spatially explicit parameters significantly enhance the spatially inexplicit assessment of ES?
- Does the identification of land use options for planning of sustainability support IWRM?

In order to support decisions for adapted land use, we used the planning support tool *Letsmap do Brasil* (Lorz et al. 2012b), developed within the GISCAME framework (Fürst et al. 2010ab20112012; Koschke et al. 20122013). As an assessment framework, we made use of the ES approach (e.g., MA2005; Haines-Young and Potschin 2010), which we intended to adopt for IWRM. We assessed the regulating ES water purification, sediment retention, and water retention and provision of food and fodder to analyze the impact of land use scenarios to mitigate pressure on water resources.

In order to allow the inclusion of site-specific conditions within the assessment, we implemented GIS-based analyses to identify areas with differing ecosystem potential (EP) into GISCAME. With EP, we mean the site-specific suitability or capacity of a natural ecosystem to provide a specific ES. Combining EP with the land use type-specific ES potential yields the hypothetical maximum provision of selected ES (Burkhard et al.: Ecosystem service potentials, flows and demands – concepts for spatial localisation, indication and quantification, in review; Burkhard et al. 2012; Bastian et al.2012).

The spatially explicit mapping and assessment of ES was called for in previous applications of the GISCAME modeling approach (Fürst et al. 2013; Koschke et al. 20122013) as it contributes to the identification of priority areas that are in need for special attention from land and resource managers and are very effective in delivering the desired services if used/managed appropriately.

METHODS

The Pipiripau River Basin

The study area is the Pipiripau river basin situated in the north-eastern part of the DF (Figure 1). The river basin area extends over 215 km^2 and is mainly covered by well-drained Ferralsols which are low in nutrients (EMBRAPA, Empresa Brasileira de Pesquisa Agropecuária (The Brazilian Agricultural Research Corporation) 1978). The Pipiripau river basin is situated within the Brazilian Central Plateau. Its altitude ranges from 920 to 1,230 m asl. The study region is categorized as a semi-humid tropical climate. Precipitation amounts on average to 1,300 mm and occurs mainly during the rainy season from November to March (Strauch et al. 2012).

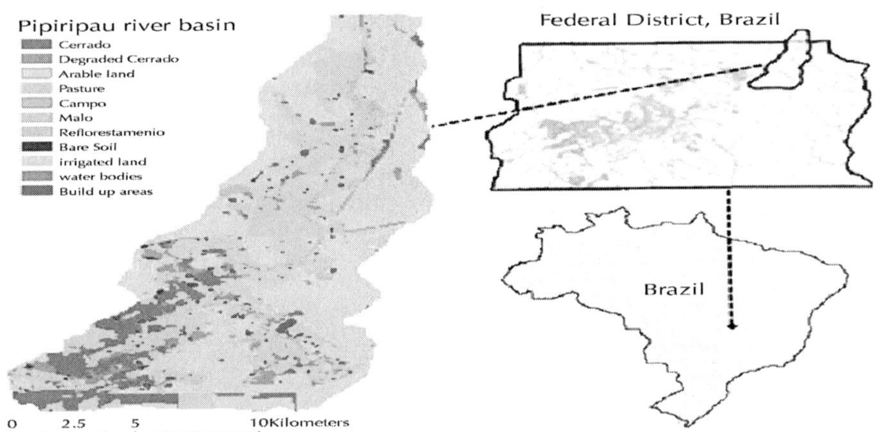

Figure 1: Location and land use/land cover of the case study area.

The basin is mostly rural, with only a few small settlements. Predominant land uses in the catchment area are arable land, which comprises foremost intensive cropping of soybean and corn (47% surface area share), and Brachiaria pasture (23%). To a smaller extent irrigated horticulture is being conducted and also remnants of Mata (natural gallery forests) can still be found along the watercourses. Mata and the different types of Cerrado, ranging from fairly dense woody Savanna of shrubs and small trees to treeless subtypes (Campo) (Oliveira-Filho and Ratter 2002), have been reduced in the last 50 years from nearly 100% to 20% of the area (Strauch et al. 2013).

Although the Pipiripau river basin is relatively small, it is the main source of drinking water for Planaltina and Sobradinho, which are satellite cities inside of the DF, and for irrigation. The river basin belongs to the project *Produtor de Água* (http://www.ana.gov.br), which aims to introduce best management practices and a better integration of stakeholders, e.g., by introducing Payments for Environmental Services (Lorz et al. 2012ab).

Assessing Ecosystem Services (ES) with GISCAME

GISCAME is a web-based software which allows a comparative assessment of ES. It consists of three components: a Geographic Information System, a Cellular Automaton, and a Multi-criteria Evaluation (Fürst et al. 2010ab). GISCAME calculates – based on raster data – the contribution of all land use types to the provision of ES on a relative scale from 0 (no provision of ES) to 100 (maximum regional provision of ES) (Fürst et al. 2010ab; Koschke et al. 20122013). The mean value of all cells of the case study area and simulated LULCC scenarios are provided as feed-back to the user. For the Pipiripau case study, the assessment of land use types vs. ES was based on the indicator values of nitrogen load (applied fertilizer in kg ha^{-1} a^{-1}) for water purification, C-Factor of the Universal Soil Loss Equation (USLE; Wischmeier and Smith 1978) for sediment retention, CN value for water retention, and yield (mg ha^{-1} a^{-1}) for provision of food and fodder. All values were standardized to the 0–100 point scale. We used the indicator values for land use types provided in Lorz et al. (2012b), IBGE, Instituto Brasileiro de Geografia e Estatística (The Brazilian National Institute of Geography and Statistics) (2006), and EMBRAPA (2012).

The presented assessment approach includes three main steps:

- Assessment of the land use type-specific contribution to ES
- Estimation of the EP (e.g., nitrogen retention potential, yield potential) based on the site-specific ecosystem properties (cf. Bastian et al. 2012)
- Combination of land use and EP maps (Figure 2)

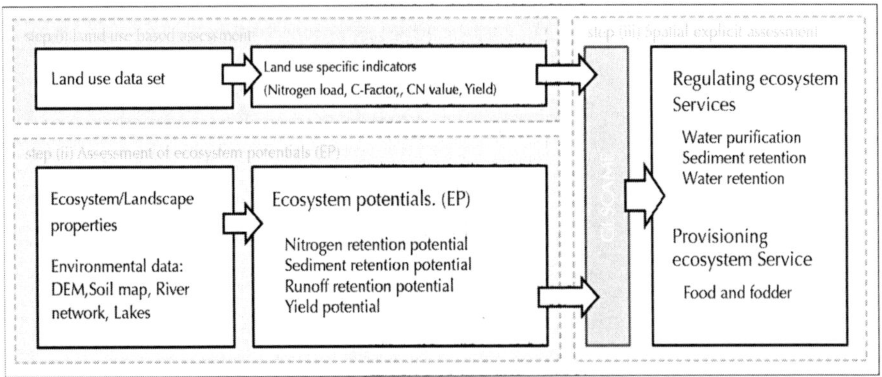

Figure 2: Assessment components and general workflow. Trough combination of land use (step i) with ecosystem potentials (EP, step ii) in GISCAME (step iii), ES can be assessed in a spatially explicit manner. Land use is the driving force which can be changed within GISCAME. Adapted from Haines-Young and Potschin (2010) and Bastian et al. (2012).

Thus, besides the land use map which is the primary information layer, a second layer that relates EP as a function of site-specific conditions to each ES was introduced in GISCAME. The secondary layer enables a spatial explicit assessment of ES by taking into account the cell-specific potential to provide individual ES. In order to couple the land use related assessment of ES (step i) with EP (step ii), we produced, for each of the ESs, a respective layer showing the related EP. According to the indicators used to assess ES we investigated the nitrogen retention potential (for water purification), the sediment retention potential (for sediment retention), the runoff retention potential (for water retention), and the yield potential (for provision of food and fodder). For the different ESs, individual assessment approaches were applied to identify EPs. The methodical approach has been operationalized already for sediment retention and water retention (Lorz et al. 2012b)

(Figure 3b,c). We therefore focus on water purification and food and fodder provision and explain the methodical approaches we applied to derive maps on the nitrogen retention and yield potential.

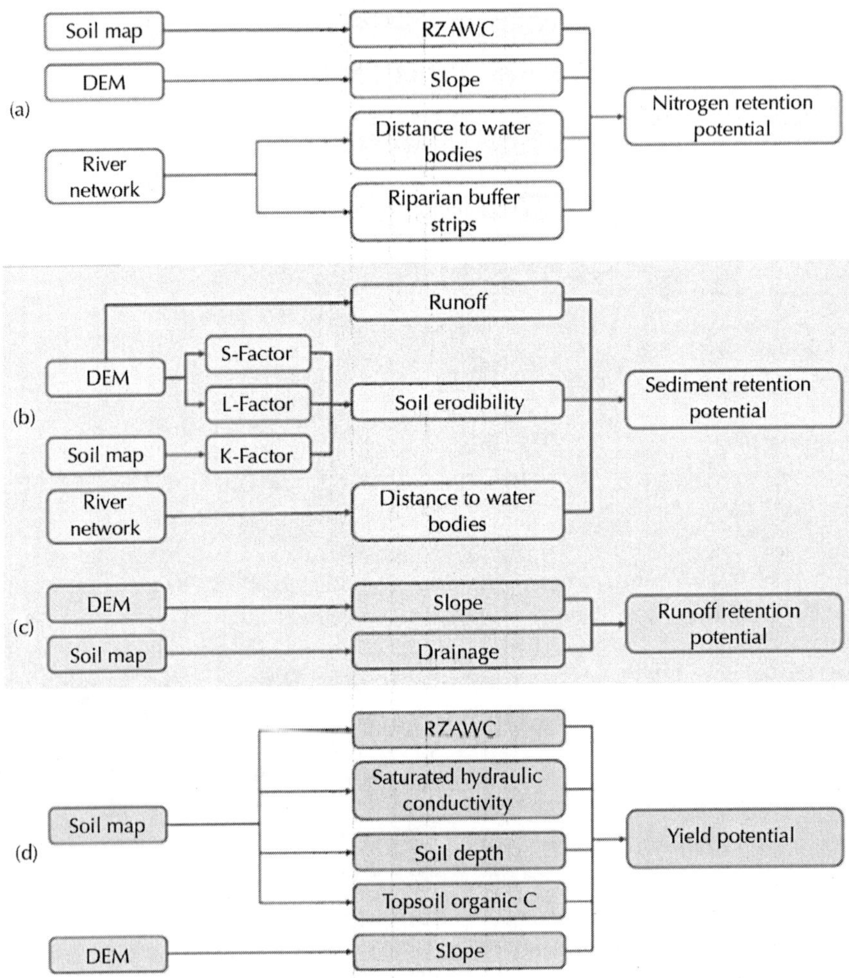

Figure 3: Scheme for the calculation of the ecosystem potentials (EPs). (a) Nitrogen retention potential (Orlikowski et al. 2011). (b)Sediment retention potential (Halbfass and Grunewald 2008). (c) Runoff retention potential (USDA, 2004), and (d) yield potential (Baja et al. 2001; Bartolini and Borselli 2009; Mueller et al. 2010). Grey background indicates that (b) and (c) were implemented in Lorz et al. (2012b).

Ecosystem Potentials (EP)

Biophysical structures and processes influence the potential of a landscape to retain nitrogen (NO_3^-) and to support provisioning of food/fodder from arable land. To assess the EP for nitrogen retention, we applied an approach modified from Orlikowski et al. (2011). The approach has been developed for river basins with low data availability and differs from other approaches (Terrado et al. 2013) in the way that it allows a qualitative estimation of the EP. We assessed the yield potential in a similar manner and developed an approach based on the available data. For soil parameters, we used data of 16 soil profiles which were the basis for a regionalization to the study area (EMBRAPA, 2012; Tomasella and Hodnett 2004). We utilized ArcMap10 to compile the data and to finally classify the parameters as well as the resulting EP maps into the three classes (i) high, (ii) intermediate, and (iii) low potential for nitrogen retention and yield production, respectively. A cell with low nitrogen retention potential would therefore exhibit a strong need for adapted land use in order to support water purification.

Nitrogen Retention Potential

Required input parameters are land use, soil (root zone available water capacity (root zone available water capacity, RZAWC)), slope, riparian buffer strips, and distance to surface waters (Table 1, Figure 3a). Our method differs from the one presented in Orlikowski et al. (2011) as follows: in order to be consistent with the previous approach of Lorz et al. (2012b) and because of the rather large scale of soil data, we used a raster cell size of 200 m × 200 m; Orlikowski et al. (2011) produced maps referring to sub-catchment level whereas we produced cell-based values. The land use parameter was not used within the data overlay. Instead, land use was assessed separately (see step i, Figure 4a) and subsequently combined with EP maps (step ii, Figure 4b) in GISCAME. Further, classes 1–3 were assigned using individual value ranges observed in the study area and not the potential overall range of values of each parameter. The preparation of input parameters is described below.

Table 1: Input data that have been used to conduct step (ii) of the assessment approach

Data	Source	Description
Land use	BRASIL 2010	Land use map of the Pipiripau river basin generated for program Produtor de Água
Digital Elevation Model (DEM)	, 1999	20 m resolution grid derived from contour line map 1:10,000
Soil	, 1978; Reatto et al., 2004; Strauch et al.2013	Soil map 1:100,000 and horizon-specific soil properties for each soil type

Koschke *et al.*

Koschke *et al. Ecological Processes* 2014 3:2 doi:10.1186/2192-1709-3-2

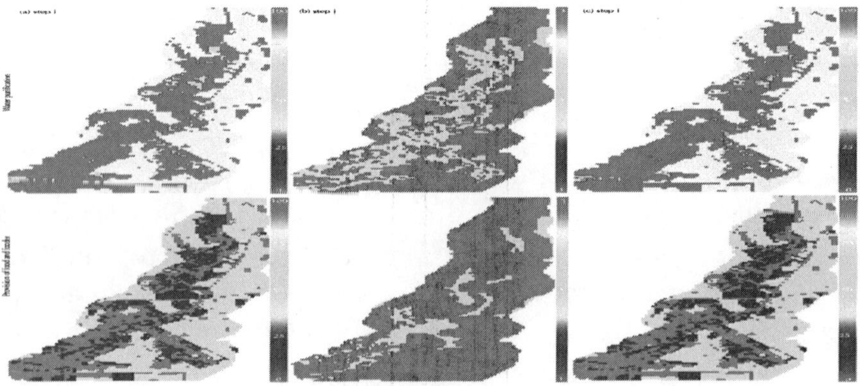

Figure 4: Resulting maps for spatial explicit assessment of water purification (above) and provision of food and fodder (below). Maps of land use based assessment (a, step i), nitrogen retention potential and yield potential (b, step ii), and combined, spatial explicit assessment (c, step iii) for the current land use pattern (Business as usual, BAU). Values in(a) and (c) range between 0–100 points, (b) comprises classes 1 (high EP, pink), 2 (intermediate, green), and 3 (low, red).

Soil: Root Zone Available Water Capacity (RZAWC)

The rationale for selecting RZAWC refers to the fact that the leaching of NO_3^- from the subsoil is commonly the main pathway to surface water bodies (e.g. Gächter et al. 2004). The loss of nitrogen is therefore influenced not only by the applied amount of fertilizer but also by soil properties. We followed the assumption that, with increasing RZAWC, the retention of water by the vegetation increases as well, leading to a lower amount of dissolved NO_3^- reaching the groundwater. We calculated the $RZAWC_{total}$ values (in mm H_2O) for the entire soil profile. We defined class thresholds on the basis of the first and third quartile: >244 mm (class 1) with mainly (very) clayey and deep Latosols and Cambisols, 142–244 mm (class 2) with intermediate properties concerning soil depth and texture and including various soil types, <142 mm (class 3) with prevailing sandy to rocky and (very) shallow Cambisols and hydromorphic soils.

Slope

Although subsurface loss of NO_3^- is supposed to be the predominant process, transport through surface runoff, which is driven by slope steepness, might occur as well (Brunet et al. 2008; Mayer et al. 2005; Orlikowski et al. 2011). We have chosen a classification of slope different from the one proposed in Orlikowski et al. (2011) in order to adapt to regional characteristics. The slope was calculated from the DEM as proposed for the S-factor of the USLE. Classes were defined according to natural breaks given in ArcMap10: 0° to 4.26° (class 1), 4.27° to 12.99° (class 2), and 13.0° to 60.43° (class 3).

Riparian Buffer Strips

Land use neighboring surface waters is important for how much NO_3^- can actually enter the water bodies. Less intensive, near-to-nature land uses such as natural vegetation, forests, and pasture in the immediate vicinity inhibit input of NO_3^- (and other nutrients) while intensively managed arable land tends to augment input of nutrients (Mayer et al. 2005).

Again, we had to deviate from the original approach because of the given resolution. Instead of the 25 m width of riparian buffer strips applied by Orlikowski et al. (2011), we used the minimum width possible, i.e., 200 m. Evaluation of land uses in terms of their ability to remove NO_3^- was conducted in accordance with Orlikowski et al. (2011). Hence, land uses Cerrado (tree Savanna), Mata (natural forest), and afforestation have the highest potential to retain NO_3^- (class 1), degraded Cerrado, pasture and meadows, and Campo (grass savanna) have an intermediate potential (class 2), and arable land (general), irrigated land, bare soil, and built up areas have a low potential (class 3, no buffer strip).

Distance to Surface Waters

The definition of classes had to be adapted to raster cell size as well. After having conducted "Euclidean distance" calculation in ArcGIS the distance to surface waters was divided into >800 m (class 1), 200–800 m (class 2), and <200 m (class 3).

Yield Potential

To detect main potential constraints for agricultural production such as water availability, water logging, and nutrient availability, we identified five main parameters (RZAWC, saturated hydrological conductivity (Ks), soil depth, topsoil organic carbon content, and slope) which are commonly used for soil productivity classification at national or regional level (e.g., Baja et al. 2001; Bartolini and Borselli2009; Mueller et al. 2010).

RZAWC

The water which is available for uptake by plants is a major factor for plant growth. We took the $RZAWC_{total}$ values (in mm H_2O) and applied the same classes as for the estimation of the nitrogen retention potential.

Saturated Hydrological Conductivity (Ks)

We calculated the mean saturated hydrological conductivity for the representative soil profiles in order to account for drainage conditions and potential water logging problems. Also, low Ks values are likely to cause runoff and induce a decreased amount of available water for plant growth (Verhoef and Egea 2012). We defined classes as follows: >42 mm h^{-1} (class 1), 15–42 mm h^{-1} (class 2), and<15 mm h^{-1} (class 3).

Soil Depth

Soil depth defines the rooting depth and the volume of soil available for water and nutrient uptake. While shallow soils tend to constrain plant growth, deeply weathered soils generally provide better growing conditions for agriculture (Blume et al. 2002). Applied soil depth classes were: >1,500 mm (class 1), 500–1,000 mm (class 2), and<500 mm (class 3).

Topsoil Organic Carbon Content

Organic carbon in soil has various effects on soil fertility both with respect to physical as well as to chemical properties. In tropical soils, organic carbon has an even greater importance as these soils are characterized through a generally low cation exchange capacity (Blume et al. 2002). A high content of organic carbon in the upper soil is therefore assumed to indicate increased nutrient availability and favorable soil structure. We classified average carbon content of uppermost soil horizons in >2.8% (class 1), 1–2.8% (class 2), and <1% (class 3).

Slope

Topography can affect productivity by means of loss of topsoil layers and nutrients through runoff. Over time, fertility and productivity of soil can be reduced. We have chosen slope classes according to Ongaro and Sarfatti (2009), as follows: <8% (class 1), 8–16% (class 2), and>16% (class 3).

We further processed the four maps related to the nitrogen retention potential and the five maps concerning the yield potential in ArcGIS10 and conducted a map overlay applying equal weights. That is, we calculated the mean class for every cell and assigned values to high (class 1, with mean values from 1.0 to 1.6), intermediate (class 2, 1.7 to 2.3), and low potential (class 3, 2.4 to 3.0). EP maps that result from step ii show the nitrogen retention potential and the yield potential (Figure 4b), or otherwise the sediment retention potential, and the runoff retention potential.

Combination of Land Use and EP

For linking EP and land use (step iii) we uploaded the maps for EP and land use into GISCAME, where the combination of both layers was achieved by percentage reductions of the initial land use-based values (Figure 4c). Percentage reductions are a result of discussion within the working group. In general, cells with a low EP are therefore subject to a higher reduction of value points than cells with a high EP (Table 2a,b). The highest reduction appears if, for example, a land use with a high fertilizer input and thus with a low value for water purification is to be found in a cell with a low potential to retain nitrogen. In terms of the yield potential, reduction was effectuated equally for all arable land uses (including pasture) as we assumed a similar effect on yield for reasons of simplicity.

Table 2: Reference matrix for the percentage reduction of land use type-related value points with respect to (a) the nitrogen retention potential and (b) the yield potential

(a)				
Land use-specific value points		Nitrogen retention potential		
	Value classes	High	Intermediate	Low
Land use types		1	2	3
Coffee	0–10	0	−10	−20
-	11–20	0	−10	−20
Arable land, cotton;	21–30	0	−10	−20
Irrigated land	31–40	0	−5	−10
-	41–50	0	−5	−10

Vegetables; Fruits	51–60	0	−5	−10
Arable land, corn tillage/no tillage	61–70	0	0	−5
Arable land, general, no tillage	71–80	0	0	−5
Arable land, bean	81–90	0	0	−5
Cerrado; Campo; Arable land, sorghum; tillage/ no tillage	91–100	0	0	0
(b)				
Land use-specific value points		Yield potential		
	Value class	High	Intermediate	Low
Land use types		1	2	3
All arable land use types	0–100	0	−10	−20

The figures in the right column refer to the percentage (%) reduction of initial value points (e.g. -10%).

Koschke *et al.*

Koschke *et al. Ecological Processes* 2014 3:2 doi:10.1186/2192-1709-3-2

LULCC Scenarios

We developed and simulated LULCC strategies and assessed their effects on the basis of the EP maps developed above in order to improve the provision of hydrological ES. Thus, LULCC scenarios (Table 3a) were driven by thematic input layers, e.g., areas with low/intermediate nitrogen retention potential and hence with high/intermediate risk for nitrogen loss, areas with high and/or intermediate slope steepness, a low distance to surface waters, etc. Further, scenarios were based on current land use, e.g., change of degraded Cerrado into Cerrado, pasture or arable land. We up-loaded the different input layers in GISCAME to simulate LULCC at the respective sites. We carried out conservation and intensification scenarios, meaning, for instance, change of land use (e.g., at low potential nitrogen retention areas) toward Cerrado, Campo, Mata or afforestation and ongoing degradation of remaining Campo and Cerrado through change toward pasture, and agriculture, respectively. The current (2010) land use pattern was used as the reference scenario (Business as usual, BAU).

Table 3: Overview and description of applied LULCC scenarios (a) and scenario results (b) according to land use-based assessment (LU) and combined land use-based and ecosystem potential (EP) assessment (LU + EP)

(a)	(b)									
		Water purification	Sediment retention			Water retention		Food and fodder		Affected area of river basin
ID	Description	LU	LU + EP	LU	LU + EP	LU	LU + EP	LU	LU + EP	[%]
BAU	Current (2010) land use distribution (baseline scenario)	88	88	86	83	48	46	31	30	-
	Change of riparian areas (Ri) with a low potential for nitrogen retention									
Ri-1	… toward native tree Savanna (Cerrado)	88	88	86	83	49	47	30	30	2
Ri-2	… toward forests (afforestation)	88	88	86	83	49	47	30	30	2
Ri-3	… toward pasture	88	88	86	83	48	46	31	30	2
Ri-4	… toward natural forest (Mata)	88	88	86	83	49	47	30	30	2
Ri-5	… toward arable land (general, no-till)	88	88	86	83	48	46	31	31	2

	Change of native grass Savanna(Campo, Ca)									
Ca-1	… toward irrigated land	86	86	85	82	46	44	34	33	3
Ca-2	… toward arable land (general, no-till)	87	87	86	82	46	44	32	32	3
Ca-3	… toward arable land (corn, till)	87	87	84	81	45	44	33	33	3
Ca-4	… toward pasture	88	88	86	82	47	45	31	31	3
	Change of areas with low nitrogen retention potential(NRP)									
NRP-1	… toward Cerrado	88	88	87	83	49	47	30	30	4
NRP-2	… toward forests (afforestation)	88	88	86	83	48	46	30	30	4
NRP-3	… toward pasture	88	88	86	83	48	46	31	30	4
NRP-4	… toward arable land (general, no-till)	88	88	86	82	46	44	32	32	4
NRP-5	… toward irrigated land	86	86	85	82	46	44	34	33	4

	Change of areas with steep slopes(SI)									
SI-1	… toward Cerrado	88	88	87	83	49	47	30	30	5
SI-2	… toward forests (afforestation)	88	88	86	83	48	46	30	30	5
SI-3	… toward pasture	88	88	86	82	47	45	31	31	5
SI-4	… toward arable land (general, no-till)	87	87	86	82	45	44	33	32	5
SI-5	… toward arable land (corn, till)	86	86	83	80	45	43	34	33	5
SI-6	… toward irrigated land	85	85	85	81	45	44	35	35	5
	Change of degraded Cerrado(DC)									
DC-1	… toward Cerrado	88	88	87	84	52	50	31	30	6
DC-2	… toward forests (afforestation)	88	83	87	83	50	48	31	30	6
DC-3	… toward natural forest (Mata)	88	88	87	84	52	50	31	30	6
DC-4	… toward arable land (general, no-till)	87	87	86	82	47	45	34	34	6

DC-5	… toward arable land (corn, no-till)	86	86	83	79	46	44	35	35	6
DC-6	… toward pasture	88	88	86	82	49	47	32	31	6
DC-7	… toward irrigated land	84	84	85	81	47	45	37	37	6
	Change of riparian areas (Ri) with an intermediate potential retention function									
Ri-6	… toward Cerrado	89	89	87	84	51	49	29	29	8
Ri-7	… toward forests (afforestation)	89	89	87	83	49	47	29	29	8
Ri-8	… toward pasture	89	89	86	83	48	46	31	30	8
Ri-9	… toward natural forest (Mata)	89	89	87	84	51	49	29	29	8
Ri-10	… toward arable land (general, no-till)	87	87	86	82	45	43	33	33	8
	Change of riparian areas (Ri) with an intermediate and low potential retention function									
Ri-11	… toward natural Cerrado	89	89	87	84	52	50	29	28	10
Ri-12	… toward forests (afforestation)	89	89	87	84	49	47	29	28	10

Ri-13	… toward pasture	89	89	86	83	48	46	30	30	10
Ri-14	… toward natural forest (Mata)	89	89	87	84	52	50	29	28	10
Ri-15	… toward arable land (general, no-till)	87	87	86	82	45	43	34	34	10
Ri-16	… toward irrigated land	82	82	84	81	45	43	39	38	10
	Change of Cerrado(Ce)									
Ce-1	… toward irrigated land	81	81	83	79	41	39	41	40	10
Ce-2	… toward arable land (general, no-till)	86	86	85	81	41	39	36	35	10
Ce-3	… toward arable land (corn, till)	84	84	80	76	39	37	38	37	10
Ce-4	… toward degraded Cerrado	88	88	85	82	42	40	31	30	10
	Distance to water (D): Change of areas close to surface waters(irrespective of their current land use)									
D-1	… toward Cerrado	89	89	88	85	53	51	28	28	14
D-2	… toward forests (afforestation)	89	89	87	84	49	47	28	28	14

D-3	… toward pasture	89	89	86	82	47	45	30	30	14
D-4	… toward arable land (general, no-till)	86	86	86	82	43	41	35	35	14
	Change of degraded Cerrado and Cerrado (DCC)									
DCC-1	… toward arable land (general, no-till)	85	85	85	81	40	38	39	38	17
DCC-2	… toward arable land (corn, no-till)	82	82	85	81	40	38	43	42	17
DCC-3	… toward pasture	88	88	84	81	45	43	33	33	17
DCC-4	… irrigated land	77	77	82	78	40	38	47	46	17
	Change of pastureland (Pa)									
Pa-1	… toward Cerrado	88	88	89	87	57	55	27	27	21
Pa-2	… toward grass Savanna (Campo)	88	88	89	87	57	55	27	27	21
Pa-3	… toward forests (afforestation)	88	88	89	85	51	49	27	27	21

Pa-4	… toward natural forest (Mata)	88	88	90	87	57	55	27	27	21
Pa-5	… arable land (general, no-till)	84	84	87	83	41	39	38	37	21
Pa-6	… toward arable land (corn, no-till)	81	81	88	84	41	39	43	42	21
	Combined (C): Change of areas with low potential for nitrogen retention, runoff retention, sediment retention, and production potential									
C-1	… toward Cerrado	90	90	89	86	56	54	27	27	23
	Change of arable land (general) (Ar)									
Ar-1	… toward arable land (soy, till)	95	95	76	73	40	39	33	32	47
Ar-2	… toward arable land (wheat)	94	94	81	78	48	46	24	23	47
Ar-3	… toward arable land (corn, no-till)	81	81	88	85	48	46	42	42	47
Ar-4	… toward Cerrado	97	97	92	90	82	81	7	7	47

(Scenarios were tabulated according to the affected area).

Koschke *et al.*

Koschke *et al.* *Ecological Processes* 2014 3:2 doi:10.1186/2192-1709-3-2

RESULTS

The maps (Figure 4a,b) were processed to qualitatively estimate the spatial explicit provision of ES. Figure 4a displays the land use-specific water purification and food and fodder provision. The colors reflect the values of the respective land uses from low (pink) to high (red). The maps that result from the GIS-based identification of EPs are depicted in Figure 4b. Figure 4c shows the results of the combined assessment for the current land use pattern (BAU). Reductions of value points only appeared in areas with a low potential to provide ES. However, the area weighted mean of water purification calculated for the whole river basin was not affected by this. This is not true for sediment retention, water retention, and food and fodder provision where the values were reduced by 3, 2, and 1 point(s), respectively, for BAU (Table 3b).

The results of LULCC scenarios are depicted in Table 3b. Compared to BAU, LULCC impact on water purification ranged from 0 to 18 value points for both the land use-based assessment and the combined assessment. Values of food and fodder provision showed a range of 23 value points throughout all scenarios and therefore had themost pronounced response to LULCC. In general, in comparison to BAU, effects grew more distinct with increasing number of altered cells. In fact, according to water purification, the assessment of the nitrogen retention potential had no effect at all compared to the spatial inexplicit land use-based assessment. As for sediment retention and water retention, the combined assessment led to reductions of 3.4 and 1.9 value points on average, respectively, compared to the land use-based assessment. For food and fodder provision, the impact of the yield potential map was also negligible with 0.4 points average reduction.

In order to go into detail we will focus on results of scenarios NRP-1, D-1, DCC-4, and C-1 (Figure 5a–d), to display the impact of more intensive and less intensive land use on the provision of ES. Referring to the combined assessment (Table 3, LU + EP), water purification was not influenced in scenario NRP-1. The change of areas close to surface waters (D-1) and the combined change of areas with a low potential to provide hydrological ES (C-1) led to a slight increase (+1 and +2 points). The change from Cerrado and degraded Cerrado into irrigated agriculture (DCC-4) led to a strong negative impact (−11). The response of sediment retention was negligible with respect to NRP-1 (+/− 0).

The increased surface area share of irrigated land in DCC-4 negatively affected sediment retention (−5) while through the change of areas near to surface waters (D-1) and through the change of low potential areas into Cerrado (C-1), an increase of 2 and 3 points, respectively, could be achieved. The performance of water retention could be increased in NRP-1, NRP-4, and C-1 by +1, +5, and +8 points, respectively. A reduction was again caused by DCC-4 (−8 points). Food and fodder provision remained the same for NRP-1. An increase of 16 points could be observed through extension of irrigated land (DCC-4). Conversely, fostering hydrological ES in scenarios D-1 and C-1 through extension of less intensive land uses reduced food and fodder provision by 2 and 3 points, respectively.

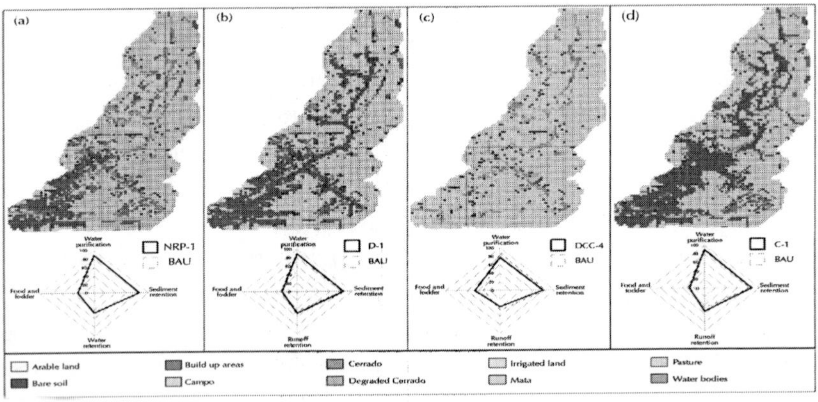

Figure 5: Land use patterns and assessment results for selected scenarios. (a) NRP-1, change of areas with low nitrogen retention potential toward Cerrado; (b) D-1, change of areas close to surface waters (irrespective of their current land use) toward Cerrado; (c) DCC-4, change of degraded Cerrado and Cerrado to arable land (general, no-till) toward irrigated land; (d) C-1, change of areas with low potential for nitrogen retention, runoff retention, sediment retention, and production potential. Resulting spider charts display scenario results (black line) and results of the initial pattern (BAU, dotted line). The different colors in the maps represent the individual land use classes (see legend below). (For interpretation of the references to color in this figure legend, the reader is referred to the web version of the article).

DISCUSSION

Assessment Approach

The underlying evaluation matrix of land uses (step i; Lorz et al. 2012b) and the procedure to produce EP maps (step ii) are subject to uncertainty and depend heavily on data quality, data availability, and resolution. In order to prevent false precision, we kept the 200 m resolution used in the setup of the system for this Brazilian case study. LULCC might have unforeseen (nonlinear) and ambiguous effects and land use-related indicator values might show widely varying ranges (Koschke et al. 2013). It is also difficult to establish a quantitative link between ES properties and potentials and ES provision (e.g., how much the nitrogen retention under arable land increases/decreases at sites with low nitrogen retention potential). Thus, the combination of the results of step i and step ii (in step iii) is based on subjective percentage reductions of value points (Table 2), but can be easily adapted, for instance, through stakeholder or expert knowledge.

In general, the approach of Orlikowski et al. (2011) to identify areas with a low, intermediate, and high nitrogen retention potential could be transferred, although some changes, for example as to the applied thresholds, were necessary. Areas that were identified as low yield potential areas in the GIS overlay match to a great extent areas that are currently covered by Cerrado, Campo, and Mata. Although, we had to assume soil related properties to be homogenous within the soil regions – which is very unlikely – the EP map seems to reflect site-specific conditions sufficiently. Although the suitability of land for crops differs widely (Elsheikh et al. 2013), differentiation of yield potential for specific crops and forest species is missing and should be the topic of future methodical refinement.

Referring to the first research question, the inclusion of environmental data constitutes a meaningful refinement of the assessment approach. The resulting EP layers could be used for two purposes: first, a spatially explicit evaluation of ES could be achieved. Second, the layers helped to identify areas that are likely to effectively improve ES provision in the case study area when managed appropriately and can be used for scenario simulations. The ES distribution maps can be displayed in

GISCAME and show results from the combined assessment (Figure 4c). This can be used as a starting point for prioritization of areas in spatial planning and management (Egoh et al. 2008; Chan et al. 2006). In the context of the case study region, more specific data might be needed.

As to the second question raised in the introduction, although the effects of the combined assessment were marginal in many scenarios in the Pipiripau river basin, the introduction of the EP and the resulting layers enhanced the previous, spatial inexplicit assessment approach (Koschke et al.20122013) considerably. It is now possible to take site-specific conditions and ES-specific processes, such as erosion and nitrogen loss, into account. Interactive training of non-experts is facilitated, for instance, to discuss the importance and challenge of considering slope, distance to surface waters, soil properties, etc., for site-adapted land use to improve the regulating ES and to reduce trade-offs with provisioning services.

A number of decision support tools have been developed to assess spatially explicit ES and in quantitative terms, e.g. InVEST (Nelson et al. 2009), ARIES (Bagstad et al. 2013), and Polyscape (Jackson et al. 2013). Although the modeling approaches of these tools tend to be more sophisticated and reliable than those presented herein, they are also much more data intensive. To our knowledge, none of these tools is able to provide feedback in real-time to end-users upon LULCC scenario simulations. We are aware of the significant simplifications we based our analyses on. However, the parameterization of EPs was based on methods used in meso-scale assessments of landscapes and is therefore an appropriate and pragmatic approach in terms of the concerned scale and the spatial and thematic resolution of data. From discussion within the working group, we concluded that the areas identified in the course of the GIS procedures to estimate EPs for water purification, sediment retention, water retention, and provision of food and fodder are reasonable. In situations where data access/availability is restricted or resources to process or collect data are not obtainable, the presented approach of using proxies, e.g., for nitrogen export such as distance to rivers and soil texture, may serve to conduct a first, robust estimation of regional potentials and potential threats.

A comparison of outcomes from studies with comparable approaches and methods and/or comparison against results of process modeling are of great importance to study uncertainty and potential errors. Nevertheless, such comparisons are difficult and could up to now not be accomplished.

LULCC Scenarios

Differences between scenarios and BAU were often marginal, especially with respect to water purification. This is because of the little area affected by LULCC in most scenarios and due to the fact that on areas with the lowest nitrogen retention potential, which are located in particular in the south-western part of the catchment area near to the drainage, often already favorable land uses such as Cerrado and degraded Cerrado are dominating (see NRP-1 to NRP-5). Thus, in the southern part, Cerrado vegetation coincides with very steep terrain. Otherwise, pasture can be found commonly on steep slopes. Arable land that extends all the way to the surface waters occurs relatively seldom. Along the river network Mata can still be found. Thus, the current land use pattern positively influences the assessment as the slope is an important factor for the considered hydrological ES. Considerable percentage reductions within the assessment procedure (Figure 4c) appear only if adverse land use coincides with areas of a low potential, e.g., areas with the prevalence of steep slopes. Since both Cerrado and pasture perform rather well in terms of water purification, and sediment- and water retention compared to arable land, the impact of EP maps in the combined assessment was often marginal as well. This was brought about by the small area with low nitrogen retention potential (4%) and the rather conservative reductions of value points we have assigned (Table 2). The impact of EP layers on the assessment in terms of sediment retention, water retention, and food and fodder provision was generally more distinct. This was due to a higher amount of adverse land uses in low-EP areas and a larger number of cells affected by percentage reductions, i.e., cells with a low and intermediate EP. Accordingly, the percentage reduction affected 16% of the area in the case of food and fodder provision, 33% as to water purification, 56% as to water retention, and 58% as to sediment retention. Consequently, the only ES affected by all four exemplary scenarios was water retention. Thus, the dependence of assessment results from basic, land use-related evaluation of LULCC was not alleviated as expected by introducing EP for water purification. This is due to the given environmental conditions, (favorable) current land use pattern, and the resulting already good performance of water purification and sediment retention.

Positive effects of LULCC scenarios were found especially for sediment and water retention. In general, less intensive land use tended to cause increased provision of water purification, and sediment- and water retention while more intensive land use such as irrigated agriculture, pasture, and arable land (Table 3, e.g., Ca-2, Ca-3, Sl-5) had contrary effects. The response of hydrological ES can be evaluated as being unidirectional; water purification, sediment retention, and water retention are affected by similar biophysical processes. Through the utilization of similar parameters to identify areas of low, intermediate, and high potential to retain nitrogen, sediments, and runoff, synergetic effects of conducted LULCC scenarios could be observed for the hydrological ES included in this study. In general, an increase of hydrological ES led to decreased provision of food and fodder (e.g., Pa-4, C-1) and vice versa (e.g., DCC-4, Pa-5). D-1 and C-1 show the trade-offs with respect to provisioning services implied with a land use that focuses on improved water resources management. The extension of intensive agriculture on remaining Cerrado (DCC-4) would in turn impair the provision of hydrological services. The results are in agreement with findings that the regulation of ES responds simultaneously to drivers of change (Bennett et al. 2009; Kandziora et al. 2013; Kramer et al. 1997; Zedler 2003) and can therefore be assumed a meaningful result of our simplified assessment approach.

Assessment results, especially in terms of sediment retention, may not appropriately reflect the problems of degraded water resources that have been identified in other reports/studies (e.g. CAESB, Environmental Sanitation Company of the Federal District 2001; Felizola et al. 2001; Fortes et al.2007; Goncalves et al. 2013; Lorz et al. 2012ab). In the case study region, water resources might be threatened more by LULCC changes that cannot be detected by commonly available data sets because of their small spatial extent or temporal occurrence which makes their mapping challenging. Franz et al. (2014) showed that sediment input into the adjacent Lago Paranoá originates mainly from urban point sources, such as construction sites, and that agricultural sites contribute only with a small amount of sediments. Small scale erosion events and point sources of sediments, however, cannot be addressed given the available land use data and assessment approaches. Land use/management change would need to be included with higher temporal resolution, i.e., with respect to interannual changes.

The need for improved land use data is supported by scenario Ar-4, the change of all arable land into Cerrado, which resulted only in a small enhancement of regulating ES. Agricultural management (e.g., soil management, fertilizer input) can be very diverse and management measures can impact ES provision at the landscape scale heavily (Strauch et al. 2013). However, these measures cannot be accounted for in the land use data set which is the core of the assessment.

The Goncalves et al. (2013) pointed out that in the coming years, population growth and excessive use of pesticides and irrigation practices may (further) reduce the amount and quality of water resources of the case study region. With respect to question three, the scenario results can therefore play an important role in the IWRM to inform people on the effects of LULCC, to initiate discussion among land managers in participatory processes, and to find best-practice land use patterns with accepted trade-offs. However, the process of regulating ES is less tangible for people and therefore may not be acknowledged (Kandziora et al. 2013). For a trade-off analysis that is meaningful for a land manager, we have included provision of food and fodder. The inclusion of other provisioning services such as drinking water supply and their linking to Payments for Environmental Services schemes would be important to communicate the benefits people obtain directly from ES (Rodriguez et al. 2006). Thus, the impact and practical relevance of applying the tool in IWRM and planning could be further enhanced.

CONCLUSIONS

In comparison to previous applications of the software platform GISCAME (Frank et al. 2012; Fürst et al. 2013; Koschke et al. 20122013) a GIS-based spatial explicit assessment was developed for this case study. We used information on soil properties and topography to introduce EPs in the assessment approach. In this study we have demonstrated how, with a basic set of environmental data and an easy-to-use GIS approach, site-specific conditions can be accounted for in a qualitative ES assessment approach. Using the layers of areas with low EP for hydrological ES as a reference (e.g., areas with low nitrogen retention potential), LULCC scenarios were conducted in areas that bear the greatest potential for positive (and negative) LULCC impacts.

With regards to scenario assessment results, we found that a considerable increase of water purification and sediment retention is difficult to achieve with reasonable, small scale land use changes as those did not affect the value points substantially. Given the current (2010) land use pattern, the Pipiripau case study area performed rather well in terms of water purification and sediment retention. There is much potential to increase water retention and provision of food and fodder which performed least well. Conservation and intensification scenarios did pinpoint the effects of efforts to restore natural systems and of further degrading land by adverse land uses, respectively. LULCC toward pastures, Campo, Cerrado, and forest in the areas with low EP to provide ES is recommended, and constitutes minimum measures to enhance water purification and sediment and water retention. More specifically, results showed that conservation measures in IWRM should be focused on arable land that is located at steep slopes. Thus, it is most likely to effectively increase hydrological ES by reducing provision services only marginally. Outcomes of the assessment indicated mono-directional, synergetic response of regulating ES to LULCC, which is in agreement with previous findings (Bennett et al. 2009; Kandziora et al. 2013). Our findings suggest that further degradation and loss of native Campo, Cerrado, and Mata must be avoided in order to safeguard water resources provision in the river basin.

In addition, the applied methods can be transferred/adapted easily for other regions and are not restricted to river basins. There is a risk to significantly over- or underestimate LULCC-related impacts due to small scale processes which cannot be adequately accounted for yet. Therefore, the methodical approach could be significantly improved by higher spatial resolution and easier availability of input data. Further, for validation of results and more appropriate inclusion of land management practices there is a need for more comprehensive monitoring (Strauch et al. 2013) of water related parameters (e.g., water quality, sediment generation), management (e.g., fertilizer application), and land use.

In spite of the existing uncertainties, the scenario results are useful for water resource managers to develop and advocate land use strategies aiming at water and soil protection in the Pipiripau river basin. The produced maps contribute to the negotiation and designation of priority areas with decision makers in (participatory) landscape planning processes to enhance water resources management and to promote resource-saving land use. Thus, in the context of IWRM, the cost-efficiency of less intensive land use to maintain and restore ES in comparison to technical solutions can be put more into focus (Maes et al. 2012). GISCAME is able to support such decision making processes as it is a powerful tool to visualize effects of LULCC and to make them more comprehensible for land managers (Fürst et al. 2013).

AUTHORS' CONTRIBUTIONS

LK did write the main part of the manuscript and collected and processed data (in GIS) necessary for ES assessment. LK did implement and process previously prepared evaluation data and was responsible for setting up Letsmap and conducting land use change scenarios and analyses. CL provided the idea and provided input with respect to the design, the scope of the study and methodical steps. He provided data, revised the first draft of the manuscript. CL also supported analyses and interpretation of results before drafting the final version. CF provided input to the conceptual and methodical approach and the linking between ES potentials and land use. She advised on the use of Letsmap. CF critically revised the manuscript before submission. TL processed data especially with respect to nitrogen retention. He adapted the assessment approach to the peculiarities of the general assessment approach and tested the applicability of different indicators and settings and the plausibility of data. Together with LK he discussed details and specific refinements of the methodical approach. FM revised the first draft of the manuscript and made proposals to improve especially the structure and the focus of the investigation. FM helped to interpret the soil data and supported discussion of possible LULCC pathways, the impact of current land use activities and availability of land data. All authors were discussing the thresholds for evaluation of environmental data considering individual value ranges. All authors participated in the discussion on the question of how much initial values have to or can be reduced in order to account for site specific potentials to provide ES. All authors read and approved the final manuscript.

ACKNOWLEDGEMENTS

The study is part of the project IWAS-ÁGUA DF, which is funded by the German Ministry of Education and Science (BMBF) in scope of the initiative 'Excellent research and innovation in the New Länder'(FKZ: 02WM1166 and 02WM1070). We thank all institutions and our Brazilian partners who have contributed to this study by providing data, knowledge, or work force. Especially we want to thank the reviewers for their valuable comments and recommendations to improve the paper.

REFERENCES

1. Bagstad KJ, Johnson GW, Voigt B, Villa F (2013) Spatial dynamics of ecosystem service flows: A comprehensive approach to quantifying actual services. Ecosyst Serv 4:117-125
2. Baja S, Chapman DM, Dragovich D (2001) A conceptual model for assessing agricultural land suitability at a catchment level using a continuous approach in GIS. In: Proceedings of the Geospatial Information and Agriculture Conference, 17–19 July 2001, Sydney. Sydney: NSW Agriculture. http://regional.org.au/au/gia/26/828baja.htm
3. Bartolini D, Borselli L (2009) Evaluation of the hydrologic soil group (HSG) with the procedure SCS curve number. In: Costantini EAC (ed) Manual of methods for soil and land evaluation, Boca Raton FL: Science Publishers. pp 118-124
4. Bastian O, Haase D, Grunewald K (2012) Ecosystem properties, potentials and services – The EPPS conceptual framework and an urban application example. Ecol Indic 21:7-16
5. Bennett EM, Peterson GD, Gordon LJ (2009) Understanding relationships among multiple ecosystem services. Ecol Lett 12:1394-1404 |
6. Blume HP, Brümmer GW, Horn R, Kandeler E, Kögel-Knabner I, Kretzschmar R, Stahr K, Wilke BM (2002) Scheffer/Schachtschabel– Lehrbuch der Bodenkunde. Heidelberg-Berlin: Spektrum Akademischer Verlag. p 593

7. Brauman KA, Daily GC, Duarte TK, Mooney HA (2007) The nature and value of ecosystem services: an overview highlighting hydrologic services. Annu Rev Environ Resour 32:67-98
8. BRASIL (2010) Programa Produtor de Água e Relatório de Diagnóstico Socioambientàl da Bacia do Ribeirão Pipiripau. DF: Brasília. p 59
9. Brunet R-C, Astin KB, Dartiguelongue S, Brunet P (2008) The mineralisation of organic nitrogen: relationship with variations in the water-table within a floodplain of the River Adour in southwest France. Water Resour Manag 22(3):277-289
10. Bryan BA, Crossman ND (2008) Systematic regional planning for multiple objective natural resource management. J Environ Manage 88:1175-1189 |
11. Burkhard B, de Groot R, Costanza R, Seppelt R, Jörgensen SE, Potschin M (2012) Solutions for sustaining natural capital and ecosystem services. Ecol Indic 21:1-6
12. Buric B, Gault J (2011) Payment for environmental services: first global inventory of schemes provisioning water for cities. Natural Resources Management and Environment Department, Land and Water Division, Rome: FAO.
13. CAESB, Companhia de Saneamento do Distrito Federal, Environmental Sanitation Company of the Federal District (2001) Plano de Protecao Ambiental do Ribeir o Pipiripau, Diagnóstico Ambiental (Environmental Report of Pipiripau basin). 1:206 http://www.emater.df.gov.br , Accessed: 22.01.2014
14. Companhia de Planejamento do Distrito Federal (Codeplan) SistemaCartográ_co do Distrito Federal - SICAD (1999) Cartasplanialtimétricasemformatovetorial, naescala 1:10.000. Brasília.
15. Chan KMA, Shaw MR, Cameron DR, Underwood EC, Daily GC (2006) Conservation planning for ecosystem services. PLoSBiol 4:2138-2152
16. de Groot R, Alkemade R, Braat L, Hein L, Willemen L (2010) Challenges in integrating the concept of ecosystem services and values in landscape planning, management and decision making. Ecol Complex 7:260-272

17. Egoh B, Reyers B, Rouget M, Richardson DM, Le Maitre DC, van Jaarsveld AS (2008) Mapping ecosystem services for planning and management, Agriculture. Ecosyst Environ 127(1–2):135-140
18. Elsheikh R, Mohamed Shariff ARB, Amiri F, Ahmad NB, Balasundram SK, Soom MAM (2013) Agriculture Land Suitability Evaluator (ALSE): A decision and planning support tool for tropical and subtropical crops. Comput Electron Agric 93:98-110
19. EMBRAPA, Empresa Brasileira de Pesquisa Agropecuária (The Brazilian Agricultural Research Corporation) (1978) Levantamento de reconhecimento dos solos do Distrito Federal. BoletimTécnico 53. Rio de Janeiro.
20. EMBRAPA, Empresa Brasileira de Pesquisa Agropecuária (The Brazilian Agricultural Research Corporation, Martha GB, Bento de Souza Ferreira Filho J (eds) (2012) Brazilian agriculture development and changes p162 http://www.cecat.embrapa.br , Accessed: 22.01.2014
21. Felizola ER, Lago F, Galvão WS (2001) Avaliação da dinâmica da paisagem no Distrito Federal. Projeto da Reserva da Biosfera do Cerrado e Fase I. In: Anais X Simpósio Brasileiro de Sensoriamento Remoto, Foz do Iguaçu, 21–26 April2001. 1600: INPE. p 1593
22. Fortes PTFO, Oliveira GIM, Crepani E, Medeiros JSD (2007) Geoprocessamento aplicadoaoplanejamento e gestãoambientalna Área de Proteção Ambiental de Cafuringa, Distrito Federal Parte 2: processamento de dados espaciais. In: Anais XIII SimpósioBrasileiro de Sensoriamento Remoto, Florianópolis, Brasil, 21–26 April2007. Florianópolis: INPE. pp 2613-2620
23. Frank S, Fürst C, Koschke L, Makeschin F (2012) A contribution towards a transfer of the ecosystem service concept to landscape planning using landscape metrics. Ecol Indic 21:30-38
24. Franz C, Makeschin F, Weiss H, Lorz C (2014) Sediments in urban river basins: Identification of sediment sources within the Lago Paranoa catchment, Brasilia DF, Brazil – using the fingerprint approach. Sci Total Environ 466–467:513-523
25. Fürst C, König H, Pietzsch K, Ende HP, Makeschin F (2010) Pimp your landscape – a generic approach for integrating regional stakeholder needs into land use planning. EcolSoc 15(3):34

26. Fürst C, Volk M, Pietzsch K, Makeschin F (2010) Pimp your landscape: a tool for qualitative evaluation of the effects of regional planning measures on ecosystem services. Environ Manage 46(6):953-968|

27. Fürst C, Lorz C, Makeschin F (2011) Integrating land management and land-cover classes to assess impacts of land use change on ecosystem services. IntJ BiodivSciEcosystServ Manage 7(3):1-14

28. Fürst C, Pietzsch K, Witt A, Frank S, Koschke L, Makeschin F (2012) How to better consider sectoral planning information in regional development planning - example afforestation and forest conversion. J Environ Planning Manage 55(7):855-883

29. Fürst C, Frank S, Witt A, Koschke L, Makeschin F (2013) Assessment of the effects of forest land use strategies on the provision of ecosystem services at regional scale. J Environ Manage 127:96-116|

30. Gächter R, Steingruber SM, Reinhardt M, Wehrli B (2004) Nutrient transfer from soil to surface waters: differences between nitrate and phosphate. AquatSci 66:117-122

31. Goncalves T, Fischer T, Gräbe A, Kolditz O, Weiss H (2013) Groundwater flow model of the Pipiripau watershed, Federal District of Brazil. Environ Earth Sci 69:617-631

32. Halbfass S, Grunewald K (2008) Ermittlung räumlich verteilter Sediment Delivery Ratio zur Modellierung von Sedimenteinträgen in Fließgewässer im mittleren Maßstab. WasserWirtschaft 3:31-35

33. Haines-Young R, Potschin M (2010) The links between biodiversity, ecosystem services and human well-being. In: Raffaelli D, Frid C (eds) Ecosystem ecology: a new synthesis, British Ecological Society, Cambridge: Cambridge University Press. pp 110-139

34. IBGE, InstitutoBrasileiro de Geografia e Estatística (The Brazilian National Institute of Geography and Statistics) (2006) Censo Agropecuário. http://www.ibge.gov.br/ . Accessed24.07.2013

35. Jackson B, Pagella T, Sinclair F, Orellana B, Henshaw A, Reynolds B, Mcintyre N, Wheater H, Eycott A (2013) Polyscape: A GIS mapping framework providing efficient and spatially explicit landscape-scale valuation of multiple ecosystem services. Landscape Urban Planning 112:74-88

36. Kandziora M, Burkhard B, Müller F (2013) Interactions of ecosystem properties, ecosystem integrity and ecosystem service indicators – A theoretical matrix exercise. Ecol Indic 28:54-78

37. Koschke L, Fürst C, Frank S, Makeschin F (2012) A multi-criteria approach for an integrated land-cover-based assessment of ecosystem services provision to support landscape planning. Ecol Indic 21:54-66

38. Koschke L, Fürst C, Lorenz M, Witt A, Frank S, Makeschin F (2013) The integration of crop rotation and tillage practices in the assessment of ecosystem services provision at the regional scale. Ecol Indic 32:157-171

39. Kramer RA, Richter DD, Pattanayak S, Sharma NP (1997) Ecological and economic analysis of watershed protection in Eastern Madagascar. J Environ Manage 49:277-295

40. Lorz C, Abbt-Braun G, Bakker F, Borges P, Börnick H, Fortes L, Frimmel FH, Gaffron A, Hebben N, Höfer R, Makeschin F, Neder K, Roig HL, Steiniger B, Strauch M, Walde DH, Weiß H, Worch E, Wummel J (2012) Challenges of an integrated water resource management for the Distrito Federal, Western Central Brazil: climate, land-use and water resources. Environ Earth Sci 65:1575-1586

41. Lorz C, Neumann C, Bakker F, Pietzsch K, Weiß H, Makeschin F (2012) A web-based planning support tool for sediment management in a meso-scale river basin in Western Central Brazil. J Environ Manage 127:15-23

42. MA (2005) Ecosystems and human well-being: Synthesis. A Report of the Millenium Ecosystem Assessement. Washington: Island Press.

43. Maes J, Egoh B, Willemen L, Liquete C, Vihervaara P, Schägner JP, Grizzetti B, Drakou EG, Notte AL, Zulian G, Bouraoui F, Paracchini LM, Braat L, Bidoglio G (2012) Mapping ecosystem services for policy support and decision making in the European Union. EcosystServ 1:31-39

44. Mayer PM, Reynolds SK Jr, Canfield TJ (2005) Riparian Buffer Width, Vegetative Cover, and Nitrogen Removal Effectiveness: A Review of Current Science and Regulations. Oklahoma: U.S. Environmental Protection Agency. p 27

45. Mueller L, Schindler U, Mirschel W, Shepherd TG, Ball B, Helming K, Rogasik J, Eulenstein F, Wiggering H (2010) Assessing the productivity function of soils. A review. Agron Sustain Dev 30:601-614

46. Nelson E, Mendoza G, Regetz J, Polasky S, Tallis H, Cameron DR, Chan KM, Daily GC, Goldstein J, Kareiva PM, Lonsdorf E, Naidoo R, Ricketts TH, Shaw MR (2009) Modeling multiple ecosystem services, biodiversity conservation, commodity production, and tradeoffs at landscape scales. Front Ecol Environ 7(1):4-11

47. Oliveira-Filho AT, Ratter JA (2002)Oliveira PS, Marquis RJ (eds) Vegetation physiognomies and woody flora of the Cerradobiome, New York: Columbia University Press. pp 91-120

48. Orlikowski D, Bugey A, Périllon C, Julich S, Guégain C, Soyeux E, Matzinger A (2011) Development of a GIS method to localize critical source areas of diffuse nitrate pollution. Water Sci Tech 64(4):892-898

49. Ongaro L, Sarfatti P (2009) Land evaluation for developing countries. In: Costantini EAC (ed) Manual of methods for soil and land evaluation, Boca Raton FL: Science Publishers. pp 125-136

50. Pattanayak SK, Wendland KJ (2007) Nature's care: diarrhea, watershed protection, and biodiversity conservation in Flores, Indonesia. BiodiversConserv 16(10):2801-2819

51. Pert PL, Butler JRA, Brodie JE, Bruce C, Honzak M, Kroon FJ, Metcalfe D, Mitchell D, Wong G (2010) A catchment-based approach to mapping hydrological ecosystem services using riparian habitat: a case study from the Wet Tropics, Australia. Ecol Complex 7:378-388

52. Reatto A, ÉdS M, Farias MFR, da Silva AV, de Carvalho OA (2004) Mapa Pedológico Digital e SIG Atualizado do Distrito Federal Escala 1:100.000 e uma Síntese do Texto Explicativo. Planaltina: Embrapa Cerrados. p 31

53. Rodriguez JP, Beard TD, Bennett EM, Cumming GS, Cork S, Agard J, Dobson AP, Peterson GD (2006) Trade-offs across space, time and ecosystem services. EcolSoc 11(1):28

54. Rounsevell MDA, Pedroli B, Erb KH, Gramberger M, Busck AG, Haberl H, Kristensen S, Kuemmerle T, Lavorel S, Lindner M, Lotze-Campen H, Metzger MJ, Murray-Rust D, Popp A, Pérez-

Soba M, Reenberg A, Vadineanu A, Verburg PH, Wolfslehner B (2012) Challenges for land system science. Land Use Policy 29:899-910

55. Schmidt N, Geier P, Mannschatz T, Matschullat J (2009) The Cerrado biome in central Brazil – natural ecology and threats to its diversity. Forum Geoökologie 20(2):44-50
56. Schwab GO, Fangmeier DD, Elliot WJ (1995) Soil and water management systems. New Jersey: John Wiley & Sons.
57. Strauch M, Bernhofer C, Koide S, Volk M, Lorz C, Makeschin F (2012) Using precipitation data ensemble for uncertainty analysis in SWAT streamflow simulation. J Hydrol 414–415:413-424
58. Strauch M, Lima JEFW, Volk M, Lorz C, Makeschin F (2013) The impact of Best Management Practices on simulated streamflow and sediment load in a Central Brazilian catchment. J Environ Manage 127:24-36
59. Terrado M, Acuna V, Ennaanay D, Tallis H, Sabater S (2013) Impact of climate extremes on hydrological ecosystem services in a heavily humanized Mediterranean basin. Ecol Indic 37(A):199-209
60. Tomasella J, Hodnett MG (2004) Pedotransfer functions for tropical soils. Dev Soil Sci 30:415-429
61. United States Department of Agriculture (USDA) (2004) Natural Resources Conservation Service, Part 630 Hydrology, National Engineering Handbook, Chapter 9 Hydrologic Soil-Cover Complexes.
62. Verhoef A, Egea G (2012) Soil water and its management. In: Gregory PJ, Nortcliff S (eds) Soil conditions and plant growth, New York-Chichester: Wiley-Blackwell. p 472
63. Wischmeier WH, Smith DD (1978) Predicting Rainfall Erosion Losses: A Guide toConservation Planning. US Government PrintingOffice, Washington, DC: Science and Education Administration. p 58
64. Zedler JB (2003) Wetlands at your service: reducing impacts of agriculture at the watershed scale. Front Ecol Environ 1:65-72

Chapter 4

Reconciling Pastoral Agriculture and Nature Conservation: Developing a Co-Management Approach in the English Uplands

Christopher J Short[1] and Janet Dwyer[2]

[1]Countryside and Community Research Institute, University of Gloucestershire, Oxstalls Campus, Gloucester, GL2 9HW, United Kingdom

[2]Professor of Rural Policy, Countryside and Community Research Institute, University of Gloucestershire, Oxstalls Campus, Gloucester, GL2 9HW, United Kingdom

ABSTRACT

The article assesses the influence of the Common Agricultural Policy (CAP) on the pastoral farming systems in a National Park within the south west of England and more recent attempts to use innovative and participatory techniques to reconcile pastoral farming systems with wildlife management. The paper confirms evidence that the economic sustainability of farm businesses in the UK involved in pastoral farming is reducing, and that wildlife-orientated schemes are changing traditional farming systems in a way that might not be in the long-term interest of wildlife. The data gathered raise questions about the cost-effectiveness of the current environmental approaches, which are government-run with centrally determined prescriptions relating to the natural resource and wildlife. Drawing on the concept of co-management, an alternative approach is identified and explored with both farmers and nature conservation regulators. This approach places more emphasis upon adjusting the traditional farming system of this marginal type of farming to align with the sustainable management of a fragile ecosystem.

BACKGROUND

The English uplands are characterised by open landscapes of moorland peaks and pastoral farmed valleys with small woods and a very low proportion of cultivated land. Just over three-quarters of the uplands in England are designated landscapes: either National Parks (NPs), classed as category V protected areas using the International Union for Conservation of Nature (IUCN) definition, or Areas of Outstanding Natural Beauty (AONBs), a national protected landscapes designation (Carruthers et al. 2009). With steep slopes, poor soils and a short growing season compared to lowland England, these are marginal farming areas, traditionally managed for the extensive, grass-based production of sheep, beef and some dairy farming, and their designation as European Union (EU) Less Favoured Areas (LFA) reflects this. They are areas of great significance for biodiversity – most of the higher land is classified as Natura 2000 sites, or the near-equivalent national designation of Site of Special Scientific Interest (SSSI). A long history of enclosure has left important features in the landscape, notably systems of field boundaries across the lower slopes and in the valleys, while the moorland itself remains largely undivided and much of it is registered shared (or 'common') grazing (Short, 2008). Many English upland landscapes also contain valuable archaeological remains because unlike other parcels of land, the soil here has remained undisturbed. All these elements coalesce to produce some of the most valued cultural landscapes in England, with a similar pattern across the upland areas of northern Europe as outlined by Eriksson (2011).

Government policy goals for the uplands over the past three decades have shifted away from support for the use of the land primarily to generate agricultural output and support farming, towards a greater emphasis upon these areas' high environmental and amenity value (Lowe and Ward 2007; Condliffe 2009; Dwyer 2010). This move towards a multi-objective and benefit approach is in line with a global shift towards a more holistic approach to policy that seems to be focusing around the ecosystem services approach developed as part of the Millennium Ecosystem Assessment (MEA 2005) and the widely respected UK report, National Ecosystem Assessment (NEA2011). Most upland areas within the UK also experience high levels of recreation and tourism activity. Various forms of exercise – walking, hiking, field and water sports and aerial recreation (gliding, etc.) - are common and some of these have had a major influence upon land use and management (shooting, in particular) and the rural economy. The uplands also play an essential role in providing much of the population with clean drinking water: of the total UK water supply, around 70% is sourced from the uplands (Natural England 2011; Reed et al. 2009). Upland land management also has a strong influence on river flows and flood risk. The peat soils in the uplands represent a major carbon store and offer important potential for further carbon sequestration, the South West in particular (Natural England 2010). Many uplands also offer significant potential for renewable energy generation from wind and water, in particular.

The English uplands face two significant challenges, both of which are common to marginal farming areas across the EU, as Nori and Gemini (2011) outline in relation to the more recent reforms of the Common Agricultural Policy (CAP). Firstly, farming incomes are low, and heavily dependent upon the support provided by the CAP. The second major challenge is to halt the environmental decline. In environmental terms, the English uplands have suffered declines in the extent and quality of important habitats and species since the 1940s, and particularly in the decades between 1970 and 2000.

This article firstly outlines the traditional pastoral farming system that has shaped the English upland landscape for the past 500 hundred years and then introduces the principles of co-management as it is developing in the UK. The article then assesses how the CAP and in particular the environmental schemes are impacting on wildlife. To illustrate this, the article uses findings from a recent study within Exmoor National Park in the south west of England (see Figure 1). The final section assesses the suitability of the concept of co-management as a possible approach that will reconcile differences between the pastoral system and the need to stem environmental decline in upland areas.

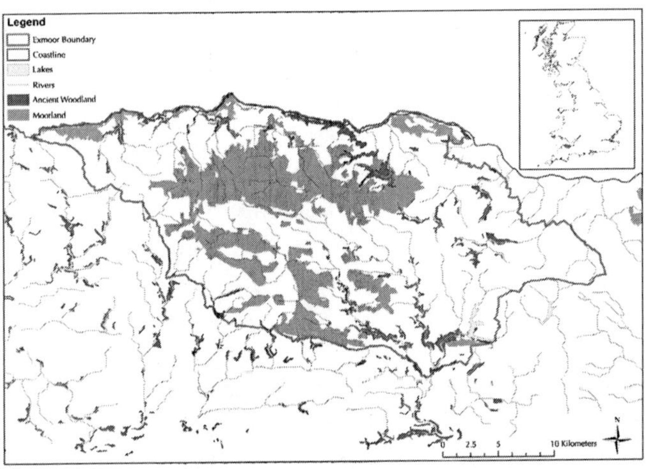

Figure 1: Exmoor National Park in the south west of England.

Overview of the English Upland Pastoral System and Study Area

Longstanding English upland farming systems comprise sheep and beef-rearing enterprises which make use of the varied contours and conditions found in these landscapes. The exposed moorland and enclosed land on the higher slopes is mainly used to rear breeding stock – hardy hill breeds of sheep and cattle, and on Exmoor, native ponies. The majority of the ewes in a hardy sheep flock are crossed with less hardy but bigger breeds to produce crossbreed flocks. Traditionally (certainly since the mid-20th century), the majority of half-hardy ewes would be sold to lowland farmers as breeding stock, and the male lambs sold as 'stores' to be fattened on lower, more productive land. If there is sufficient lower enclosed land (often called in-bye land) available, an upland farm may also run its own cross-bred flock which is more productive than the hill breeds (as measured in both lambing percentages and carcass weights).

Upland farm cattle are predominantly suckler cow herds which are used to produce beef calves. While the hardy breeding animals can graze the moor all year-round, their progeny will usually graze the hills only during the summer, and in winter they will be moved down to lower-lying pastures, prior to sale off the holding or recruitment into the breeding flock or herd, in the following year. Commonly, grazing land for these young stock was provided on upland dairy farms often located in the valley bottoms, where the milking herd was generally kept indoors through the winter.

During the 1970s and 1980s, there is evidence that a combination of market trends and CAP support payments encouraged many hill farms to increase stock numbers and the intensity of land management in these areas (Baldock et al. 2002). The market trends show a steady increase in sheep and cattle numbers, largely because the support payments were headage payments with farmers receiving payments per head for their number of breeding ewes. As (Baldock et al. 2002) point out, this strongly influenced farmer behaviour and led to an increase in the number of stock grazing upland areas. By the early 1990s the major threat to biodiversity in upland areas was overgrazing by sheep and cattle (English Nature 1996). Following the 1992 CAP reforms, cross-compliance conditions were introduced on sheep and beef support payments which were specifically designed to deter overgrazing, and during the 1990s stock numbers in these areas stabilised (Baldock et al. 2002). In the same period, environmental schemes for upland farms specifically sought to reduce overgrazing of moorland, as well as encouraging a range of positive management actions, including restoring traditional field boundaries and protecting farm woodlands from damage by livestock, through fencing. However, in 2002, the government's nature conservation agency re-stated that overgrazing was one of the key issues undermining favourable status on Natura

2000 sites in England (English Nature 2002)and it was expressed again by Natural England (NE) in 2009 (NE 2009), the English government's nature conservation agency. During this study, one of the interviewees outlined the concerns of NE that the:

'Elements of the intensive livestock model of upland farming we saw in the past were an environmental disaster (high stocking rates, wetland drainage, fertiliser applications) leading to unfavourable condition on many upland SSSIs (species and habitat losses), diffuse water pollution, soil erosion and carbon loss.' (NE adviser pers. comms. 2009)

This concern explains why many upland areas have been targeted with a variety of environmental schemes since the late 1980s, offering payments to farmers for reductions in stocking levels per hectare, notably on the moorland areas of the farm. This is matched by regulations linked to the receipt of support payments (often called cross compliance) to prevent severe overgrazing, and has been applied to agricultural policy support since 1991.

A pictorial representation of the interlinked use of the upland landscape in 'traditional' hill farming systems, as well as a summary of long-term trends in farm business change (c.1980-2005), is given in Figure 2.

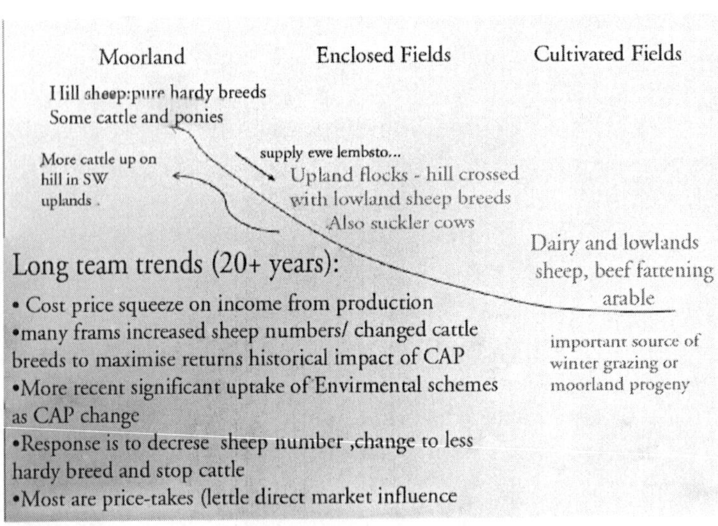

Figure 2: 'Traditional' hill farming systems and lon g-term trends in farm business change.

Farming businesses in these areas have generally limited options for farm diversification, due to their distance from urban centres and markets. Nevertheless, many receive some income from non-farming activities. Whilst upland farm businesses are noted for their resilience in the face of harsh economic and environmental conditions, they can also be particularly prone to social disadvantage, as a combined result of poor access to services (CRC 2008) and for some, poor living standards. Most of them have a very strong cultural attachment to their way of life and the traditions associated with hill farming (Defra ACEO 2009). The protection of upland farming culture, such as the traditional farming system and presence of hardy upland sheep breeds, is a highly emotive issue, as (Swales et al. 2004) reported in their evaluation of the uplands:

'The strength of feeling on many of the issues raised by our research is worthy of comment. One respondent said, 'We would stress the importance of the need to involve emotional aspects simply because to exclude them would ignore the 'heart' which has been pivotal to the structure of [these] rural areas for centuries." (p.29)

This article focuses on Exmoor and in particular the area of Exmoor National Park; see Figure 1. In Exmoor, the relationship between longstanding marginal farming practice and the environmental qualities of these landscapes is well acknowledged – as reflected in the Exmoor National Park Management Plan (2007) :

'The historical character of Exmoor is as a managed landscape, in other words there has been a relationship between man and the environment over many thousands of years. Furthermore, the flora and fauna require a mixed landscape. The significant challenge will be to work with farmers and land managers in a way that recognises that though much of the land is in commercial management, it can be sensitively managed for a variety of other benefits including high quality landscape; habitat and wildlife management; archaeology and historic buildings, and quiet enjoyment by the public. Ultimately, Exmoor National Park will only retain its special qualities with support from the people who own and manage the land.' (p.16)

Exmoor National Park supports a rich and diverse range of habitats and species. On the moorland this extends to areas of open heather and grass with upland bogs and deeply incised wooded valleys (locally called combes) blanketed in upland oak woods. Amongst the special flora and fauna are veteran trees, all species of British bats, rare lichen and bryophyte communities, notable butterfly populations, including the high brown, heath and marsh fritillary butterflies and an impressive assemblage of heathland birds. The ancient farmed landscape features small hedged fields (see Figure 3, in Exmoor these are characteristically a ditch with a bank and a beech hedge planted on top), unimproved acid grassland, narrow lanes, orchards, stone barns and farmsteads and specific breeds of livestock, namely Exmoor Horn sheep and Devon cattle.

Figure 3: An ancient farmed landscape.

The agricultural economy of Exmoor is dominated by hill farming, although on the better land to the north and east of Exmoor, arable crops are grown. The typical upland farms have access to a combination of open, rough moorland for grazing and enclosed pasture of varying quality where silage or hay may be cut and, rarely, root crops can be grown, for winter feed. In Exmoor, most farms are owned by the farmers themselves. In this sense the valued Exmoor landscape owes its existence to agriculture but is also threatened by recent changes to the traditional farming systems. Having outlined the traditional pastoral approach in the English uplands, the next section will look at the economic impact of recent changes in the industry.

METHODS

To investigate the impact of policy, both agricultural and environmental, and the combined effect on farm businesses and land management activity on the biodiversity of the uplands, 20 in-depth interviews were undertaken, each with a detailed, selective and iterative approach which would enable causal linkages between policy design, delivery and farming systems and practices to be explored at the level of individual farm businesses. This facilitated an understanding of how farm businesses are responding to different policy and non-policy drivers, as seen in the ways in which they develop their businesses and carry out their land management. In turn, such an understanding would indicate the extent to which these strategies and practices are likely to generate and sustain the environmental and other public benefits sought by government.

The in-depth interviews were supplemented by three other stages:

- Before the interviews took place a review was undertaken of specific sources of recent literature and data on the local context and policy history, including analysis of available farm survey datasets;
- Following the in-depth interviews, a workshop was held for all those interviewed and also two or three local delivery agency officers, to help substantiate findings from interviews and to explore causal linkages between policies and farm changes in more detail;
- The final stage was for feedback to be received before a refinement in the dialogue and development of policy recommendations in collaboration with those groups consulted and involved in the Exmoor.

In these terms, this research is focused upon the impact of policy as much as content; it involves deducing the impacts with the local stakeholders, in order to expose the causal linkages between policy and changes on the ground; and it is designed to enable the evaluators to be external, but working in a transparent and open way with the various groups.

The approach also reflects that used by the GEMCONBIO project, a European-based project that sought to develop new policy instruments concerning governance and management of environmental, social and economic emergencies and a process for the development of adaptation strategies (Simoncini et al. 2008:28). The GEMCONBIO project aimed to develop these guidelines using an ecosystem approach, which emphasised the need for participation and arises out of the recent Millennium Ecosystem Assessment (MEA 2005). The type of approach taken by GEMCONBIO and developed in Exmoor reflects the principles and process of co-management, as outlined by Carlsson and Berkes (2005), as 'the result of extensive deliberation and negotiation' - meaning it is very much a process rather than a fixed state. The World Bank has described co-management as a 'decentralized approach to decision-making that involves the local users in the decision-making process' (The World Bank 1999:11).

Co-management has its roots in the work of Likens, Burch and others who combined social theory with ecology to develop social-ecological systems (Olssen et al. 2004) or human ecosystems (Likens 1992). These recognise the impact of human activities over time to the extent that the nature and the social are combined and deeply connected. The applied nature of this approach and the desire to change the more polarised approach of 'nature', and 'society', has led to these concepts developed further through approaches such as 'community-based conservation' (Berkes2003) and 'adaptive management' (Jacobson et al. 2009), both, like co-management, starting from the premise that conservation and community development can be achieved in parallel. This centres on a significant shift in ecological thinking that recognises humans as part of the ecosystem and the need for participatory approaches to identify and integrate 'traditional' human activities into conservation management.

Thus the co-management concept, as outlined by Carlsson and Berkes (2005), fits with the attempted reconciliation of the pastoralist system used by farmers in the uplands and wildlife management. Carlsson and Berkes suggest six steps as a means of developing a co-management approach, which involve (adapted from Carlsson and Berkes 2005:73–4):

Defining the sociological system – i.e. identifying the group, community or system to assess and determining its various stakeholders and structure(s);

- *Mapping the essential tasks and key issues* – what activities are performed, who makes this decision and how are these decisions related and problems resolved;

- *Clarifying the participants* – who is involved, directly or indirectly in the activities and decisions identified in the mapping process;
- *Analysing linkages* – what are the connections between these activities within the area and their impact on neighbouring systems and areas;
- *Evaluating capacity needs* – a comprehensive review of the institutional arrangements and the capacity of the various stakeholders;
- *Developing a way forward* – involving all stakeholders in shaping an adaptive process that provides an integrated process for future decision-making.

They conclude that by applying such a co-management approach, underpinned by these six steps, 'power sharing will typically be regarded as the end result of a collaborative problem solving process rather than a starting point of a co-management decision-making process' (Carlsson and Berkes 2005:74). These is in stark contrast to conventional resource management approaches as outlined by (Berkes et al. 2000), which are characterised by rules and regulations, established by a central agency and enforced by agents with little connection to the resource. It is this distinction between the two approaches and the perceived notion that agriculture and wildlife policy have become polarised that suggests that co-management might provide a way forward for the uplands. The next section analyses existing datasets in order to grasp the main challenges to upland farm businesses in England, and more specifically the south west.

Analysis of Existing Data on Farm Businesses in the English Uplands

Under the 2005 reforms to the CAP, England opted to separate all direct payments; that is the support paid to farmers based on area of crops and number of livestock, a process often referred to as 'decoupling' payments from production, from 1 January 2005. In addition, whilst entitlements to this support were initially (in 2005) allocated to each farm on a historic basis (thus reflecting past productivity), the Department of Environment, Food and Rural Affairs (Defra) introduced a system which will gradually reduce the proportion of payments made on a historic basis and replace this with payments made on a standard area basis, in a phased way over eight years, up to 2013. For the area payment, there are three classes of land for which different payment rates per hectare apply:

- The lowest rate is paid on moorland, which in 2009 was around $40/ha;
- A higher payment applies on LFA land which is classified as 'Severely Disadvantaged Area' (SDA), $230/ha, and
- A uniform, higher payment applies to all other farmland in England which falls outside these two categories, including upland areas formerly designated as 'Disadvantaged Area' within the LFA, $280/ha (figures from Defra, as at November 2009).

There is widespread recognition that farming incomes in the English uplands are low, and heavily dependent upon the support secured through the CAP. The Farm Business Survey, which collects data from across the UK, shows that in 2007, the most marginal areas in England earned 50% of their total revenue (output) from crop and livestock farming activities with 42% coming in various forms of support provided by the CAP (Franks et al. 2008). The balance of revenues was earned from non-farming or diversification activities. In 2006–7 the average LFA farm was losing money on its commercial output, on average this amounted to -$25,520 in 'Farm Business Income' (meaning a loss rather than a profit before accounting for family labour). However, revenues earned from the CAP under the single farm payment, environmental schemes and diversification offset the losses in farm business Income (sale of products minus costs such as fuel and feed), to generate a net average return of $27,160. Nevertheless, this income does not include the cost of unpaid labour, provided mainly by the farm family. As with many average financial figures, there is substantial variation amongst farms in terms of their commercial performance, but it is clear that many upland farms would struggle to continue as commercial businesses without continued public support (Franks et al. 2008).

The figures for the south west uplands suggest that average farm business income is projected to fall by 34% because of the SPS changes, possibly rising to a cut of 64% for the larger 'mixed grazing livestock' farms between 2008 and 2012 (Turner et al. 2008). Under the worst case scenario the cut in income would be 43% on average and 76% for the mixed grazing livestock farm (the type of farm most likely to provide grazing on moorland). A recent Defra survey which drew responses from a large and statistically representative sample of over 1,000 upland farms across England, has provided information on how farmers have been adapting their farm management to current economic and other conditions, since the implementation of full decoupling in 2005 (Defra ACEO 2009).

In the last four years, 36% of upland farmers have reduced or stopped grazing on moorland, which is the highest most marginal land in the uplands. The most common reasons given were due to environmental schemes (61%) and the poor financial return of upland livestock (38%). These figures suggest that farmers have responded significantly to the incentives of environmental schemes to reduce overgrazing on moorland areas, but indicate that wider economic factors have also reduced the business incentive to graze moorland. While the changes in CAP support outlined above are important, consistently weak market prices up to 2007 (Gaskell et al. 2008) may also be relevant. Such reductions may also hinder farm businesses responding when market conditions improve, as they have since 2009.

Within Exmoor analysis of farm survey data (Defra 2011) over a ten-year trend (2000–09) highlights the decrease in stock numbers; especially sheep (see Figure 4).

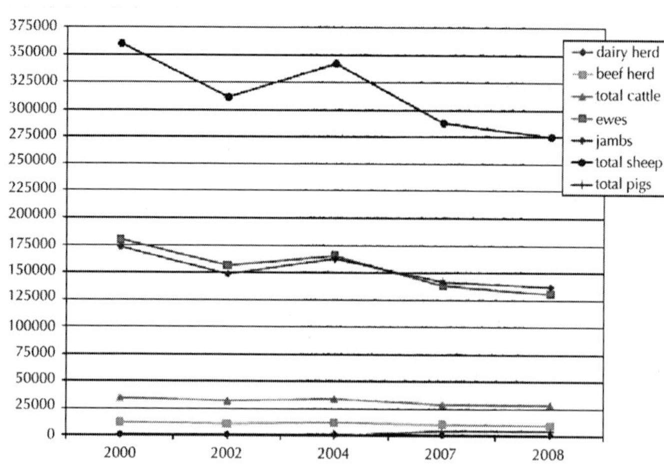

(Source Defra 2011)

Figure 4: Exmoor stock numbers from 2000 to 2008.

The table shows a reduction of about just under 10,000 in the total sheep flock across Exmoor in the 10 year period. Of these 5,000 are ewes. The slightly lower reduction in lambs points to a marginal increase in lambing percentages, which would suggest a move to less hardy and more productive breeds. The cattle numbers are already low but indicate a small reduction over the same period.

Other figures for Exmoor also indicate a shift from full-time to part-time farmers and a decline in the number of medium-sized 50–100 ha holdings, with increases in the numbers of very small and larger holdings. As farms enlarge and medium-sized units disappear, this reduces the availability of 'starter units' for new entrants. The Defra survey (Defra ACEO 2009) across the whole of England found that just over a fifth of upland farmers expect that their business will not continue beyond the next five years. These are largely farmers aged over 55 and those less likely to have a son or daughter to take over the farm business. This is within a sector of the industry and that is classically quite static, with almost two thirds of farms (64%) being long established family businesses and 25% being first generation family farms. The same survey found that on 37% of upland farm businesses succession is secured (almost entirely within the family). These figures suggest a community which is uncertain about its future viability and may be having difficulty attracting the next generation into farming.

The next section looks in detail at how farmers have been responding to these challenges and the impact of the policy mechanisms introduced to alleviate them.

How Upland Farmers are Responding to these Challenges

The 20 in-depth interviews were undertaken with a sample that include a range of farm sizes, types/systems, tenures and family/social situations that reflected Exmoor as a whole, according to the research identified through literature review. The interview used a semi-structured format with questions designed to review a ten-year period of farm changes and a more detailed evaluation of the impact of existing and past policies. At the end of the interview, views on likely future change, concerns and aspirations were gathered. The interviews all took between 1.5 and 2.5 hours, and all took place on the farm (face-to-face). This was followed by a workshop to which all of the farmers were invited and it is at this stage that a new framework reflecting the co-management approach was discussed.

The interviews revealed four distinct, but often overlapping, strategies of farm business development since 2000:

- Farm enlargement, in order to maintain or increase income; some farmers interviewed had doubled or trebled the size of their farms.
- Simplifying the farm business by ceasing one enterprise, such as dairy or beef production, often because they require more labour costs than sheep production.
- Supplementing or subsidising the farm income with income earned from non-farming or off-farm activities.
- Adjusting the farming system in response to environmental schemes and/or the introduction of decouple payments in 2005, which removed the link between farm support and stock numbers.

Enlargement was a strategy used by just over half of those interviewed; three farms doubled or trebled in size and a further eight had made more modest increases. None had decreased the area they farmed. In most cases, additional land has been bought or rented as it has come available nearby, with farmers seeing it as an investment (small increases) or as a necessity to continue farming and plan for succession. One farmer explained how when the succession on their farm was agreed about ten years ago, the family calculated that it would need about 400 hectares to support each family member, so they needed more land. The family therefore increased the farm area in 2000 (by 117 ha), 2002 (50 ha), 2004 (45 ha) and in 2008 (490 ha). They also increased stock, trebling sheep and cattle numbers. However, as some family farms expand in order to meet the needs of succession, others with no successor cease trading, thereby providing the land for such transactions. The most common result of this adjustment is for fixed costs to be spread more thinly across a larger unit, gaining economies of scale, but also stretching the management capability of the family.

Simplifying their business was a strategy of five of the 20 by ceasing one enterprise type among the mix. In the case of dairying, one had left this enterprise and turned to non-farming uses of the land while the other had decided to focus exclusively on the dairy. In other cases, labour costs are a driver: one farm ceased beef production because cattle require more labour than sheep but on another the farmer felt the reverse was the case and reduced his flock to a small pedigree flock and doubled cattle numbers. The other reasons given for enterprise changes were to meet environmental scheme requirements (discussed below) and to reflect the impact of the Single Payment Scheme, and in one case, the impact of deciding to convert to organic production, which meant the land could not carry so many livestock.

Supplementing or subsidising the farm income with income earned from non-farming or off-farm activities has been a common strategy that started before 2000. Commonly, this involves the spouse of the main farmer and/or son having a job outside farming or the main farmer having a second income. For example within the sample, a spouse has another job off the farm: some do farm recording; others work in project management, teaching and the leisure industry. In several cases, the principal farmer had another part-time job which was more closely linked to farming: two are farm contractors providing services for neighbouring farms; two others are farm business consultants offering professional advice and help with form-filling; and at least two businesses make some money from shooting. At least four farms have income from holiday accommodation including bed and breakfast, camping and renting cottages. Thus the overwhelming majority of interviewees derive a share of their total household income from sources other than the farm itself. However, this tended not to involve any of those who had significantly increased the size of the farm or specialised into particular enterprises, suggesting this absorbed the family labour on the farm.

All of the farmers had entered the farm into agri-environment schemes during the preceding 20 years. All farms had also been affected by the introduction of the Single Farm Payment, the decoupled support introduced in 2005, which, as explained earlier, removed the link between farm support and stock numbers. In response to these two factors, many farmers had adjusted the farming system. For others, a move to fewer hardy hill stock has been a response to the payment opportunities presented by agri-environment schemes as reductions in stocking levels per hectare, notably on the moorland areas of the farm, are required under the schemes. The reductions have been significant with three farmers no longer grazing the moor and a further four reducing their livestock by at least half. As one farmer put it 'having a few stock on the moor, now, it just isn't worth it – labour or stock-quality wise'.

Decoupling, and the introduction of an area-based payment that is not linked to production, has also increased the relative attractiveness of this option. In almost half of all interview cases, the consequence of this has been a significant change in the emphasis of the overall land management.

As far as grazing the moorland was concerned at least four farmers had given up their pure bred hill sheep flocks while all others are spreading their sheep flock more thinly across larger areas of land. One of the farmers noted that the population of the local bred of sheep, the pure-bred Exmoor Horns, had fallen from thousands to just hundreds in about 20 years as a result of this change. As the number of sheep grazing the moors (most of them common grazings), decreases it becomes harder to gather the livestock at key times of the year, and this is exacerbated by the reduced number of people available to undertake the task. However, while the number of potential and actual graziers has decreased there has been a revival in the social structure associated with commons, largely because the environmental schemes require a constituted committee of rightsholders. In many cases such gatherings had hardly been functioning but this requirement has led to an increased sense of social cohesion, collective management and communication between the remaining farmers.

The result of this adjustment to the farming system is that the moorland is no longer an integral part of the commercial farming system and is consequently managed minimally, and largely for what the farmers see as 'cosmetic' purposes, as required under agri-environment agreements (from a different perspective, this management is clearly for environmental purposes). The means of implementing the required management is, in some cases, the use of extensive, hardy cattle or Exmoor ponies to graze the moor, supporting the separation of this land in management terms. The impact on the lower slopes is to have more intensive management in order to carry the main body of income-generating livestock production, further fragmenting the traditional system that linked the moors and the lower slopes.

When compared to the traditional upland farming model described earlier in Figure 2 and outlined in Figure 5, the combination of management strategies described above appears to have led to an increased degree of fragmentation of farming systems across the landscape, so that the tiered system, in which the use of the hill, the in-bye and the lower more fertile valleys was functionally linked throughout the year, has broken down on many of the farms investigated.

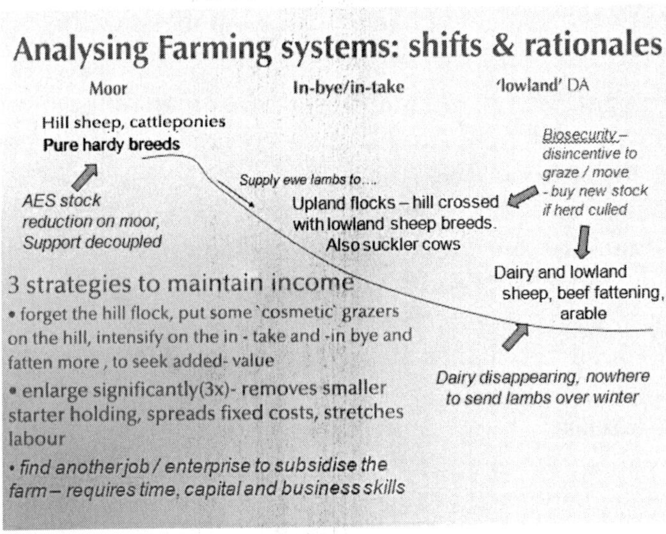

Figure 5: Traditional upland farming model and strategies to maintain income.

Impacts of Farming System Changes on Environment and Wildlife

In environmental terms these changes to the traditional farming system have produced a mix of negative and positive impacts in quite a short time. There is some improvement on the moors that had in the past been overgrazed, and the most recent data from supports this (NE 2011). However, should these lower stocking rates continue, scrub cover such as gorse, will continue to increase, raising the potential fire risk. As the lower slopes are farmed more intensively, the management differentiation between the moors and the enclosed land becomes wider. This harms the diversity and connections between different habitats that are key to many characteristic upland species, notably birds (Boatman et al. 2008).

Farmers' feelings about these changes vary. There is concern about a perceived decline in traditional hill management. There is also a feeling that this process has been pushed too far, by a combination of environmental policy and commercial factors.

With regards to the environmental schemes, the balance of opinion is sympathetic, although most farmers admit that a key reason for joining these schemes is financial and to maintain incomes. Nevertheless, there is some resentment that the top-down prescribed management is in some ways too restricted and lacks flexibility. Many claimed that whilst initial destocking had brought some benefits for heather and other upland habitat, continued management with only very low levels of sheep or cattle grazing was now leading to farmers' impoverishment. Some suggested that there was a lack of understanding that reversing business decisions about stock numbers would not be easy, for farmers. As one farmer said 'you can't keep your stock waiting on a washing line and drop them down as and when required.'

Several farmers were concerned about the prescriptions, especially stocking rates, and the impact of not meeting the requirements (minimum or maximum), noted that *'if we don't hit their targets then they might want their money back'*. Crucially they felt these issues, and the impact of the environmental scheme had not been openly discussed, as these quotes show.

The farm has become more extensive in the past 10 years, the hay quantity is down on the land in the scheme, but plant species are up: it looks rubbish to me and the stock don't like it, but NE [Natural England] love it and say it is species-rich now.

A closer working relationship between the farmer and the scheme adviser is needed. Farmers are scared of not doing things right by the prescriptions of the schemes; we also have other knowledge to offer.

I wish I could walk the fell [a local term for moor] and the areas we manage with the [environmental scheme] officer then we'd have a better and closer working relationship.

Farmers are adaptable but agencies seem to take the view 'we have the money so you have to listen to us'. A new approach is needed which is more farmer-focused.

This desire of farmers to be part of the solution in the uplands, rather than a 'service provider' for the national environmental agency, was explored further in the workshop that followed the interviews. About half of the interviewed farmers attended this workshop which was also attended by the local environmental scheme officer and the national park officer. In this setting the farmers were prepared to discuss what a new scheme might include, such as the creation of new habitats, management of landscape features, mechanisms for developing mutual skills in farming and environmental management. The framework for such a new scheme was to be based around discussions, a collective approach and a business focus that included an audit of each farm to show their biodiversity and landscape assets and potential. The emphasis here was that if the scheme could be locally developed, then they would be more willing to go further in the environmental management. As these three farmers put it:

'We are looking for one scheme which can be tailored down to individual farms, making the link with their business aims and plans.'

'The old schemes were manufactured nationally, we need a local situation.'

'A whole farm plan that covers all aspects is better than numerous separate agreements or schemes: a holistic plan that does not single out bits but sets objectives that farms can fit into and secure funding from.'

The tenor of the discussion clearly indicates a willingness among those present to work with the government agencies to develop a new approach. This would be more flexible, enabling a package of measures and commitments to be agreed for each farm which aligns the future business development goals with the environmental needs at the landscape scale. The proposal agreed by all those involved was to develop discussion groups around particular themes: biodiversity (upland moors, pastures and ecological networks), water (rivers, blanket bog and land management) and carbon (upland oak woodland, hedgerows, wet woodland and the woodland economy). Membership of each group would be made up of statutory agencies, local conservation bodies and farmer/landowners who, through a facilitated process, will develop an action plan. Through this shared learning process, each group would identify the location of areas for enhancement, connectivity and new habitat creation, and the appropriate mechanisms to achieve these, in line with the land management needs and biodiversity strategies.

Through such a co-management approach, the policy process could serve the valuable goals of building social capital, institutional trust and thus a higher commitment to achieving policy objectives, among its key beneficiaries. This would offer more potential for policy learning within the various agencies, as well as supporting the objectives and values of local stakeholders. This approach thus differs from the more regulatory and incentive-driven approach currently in operation.

CONCLUSIONS

The proposal to use a co-management approach for a framework that would deliver both agricultural and environmental benefits, seems to be a requirement, because of the multiple local interests and multiple government agencies within a relatively small area. In order to meet multiple state and community objectives, a co-management approach is beneficial in securing long-lasting governance. As Carlsson and Berkes (2005:65) note, the approach should be seen as a 'continuous problem solving process ... involving extensive deliberation, negotiation and joint learning [original emphasis]. This finding is also reinforced by Reed (2008) and Leys and Vanclay (2011) who both identified the need for a greater use of participatory techniques in environmentally diverse and agriculturally marginal areas. The continuation of farming practices, skills, customs and traditions is explicitly supported in the policy statements and visions of most UK government agencies, including the environmental agencies (NE 2009;ENPA 2007;Carruthers et al.2009).

The findings of this case study endorse a case for a new approach to economic and social regeneration in upland farming areas, working closely with the farming population as well as those other sectors and service providers with whom they interact most closely. This conclusion chimes well with other reviews, such as the Upland Communities Inquiry (CRC 2010), which has called for a more integrated and socially developed approach to policies for these areas, in future. It is worth noting that the GEMCONBIO project concludes that biodiversity conservation needs to be determined from local economic and social characteristics as well as local, national and international ecological needs. The policy recommendations include the need to 'recognise and respect customary institutions for natural resource management' and to 'foster alliances between local, traditional institutions governing natural resources and the governmental agencies in charge of conservation' (Simoncini et al. 2008). This is part of the call to recognise the localised connections between nature and society and mirrors other similar initiatives (Short et al. 2010).

This article supports the wider assimilation into policy of the policy recommendations of the GEMCONBIO project, to 'make contact with traditional structures' and involve them in both management activities and governance. However we need to be cautious about boarding what Kasperson (2006) calls the 'stakeholder involvement express' just because it is the trendiest train in the station. Ribot (2006) makes clear that stakeholder involvement has to be meaningful, since as with all decentralised approaches, they are 'only effective when there are mechanisms to represent local needs and aspirations in decision making'.

While many upland public goods and benefits are directly associated with land and land management, these are also connected with upland communities and the combination of land and people. The findings of this article point to a need for a more territorially-focused and multipurpose policy design and delivery for sustainable rural development in the English uplands (Ward and Lowe 2007). The model should offer an environmental scheme that meets the combined economic and environmental pressure of each distinctive upland area, perhaps more akin to a landscape-scale approach (Lawton 2010) which also understands the behaviours, motivations, and aspirations of farmers and landowners, which will create a synergy between pastoralism and wildlife.

AUTHORS' CONTRIBUTION

CS and JD developed the research approach and undertook the research project in Exmoor. JD wrote the research report and CS was the main author of this journal article. All authors read and approved the final manuscript.

ACKNOWLEDGEMENTS

The authors would like to thank the European Commission, Exmoor National Park Authority (ENPA) and Natural England (NE) for providing the funding for this work and acknowledge the contribution made by Ian Condliffe in undertaking some of the interviews. The research would not have been possible without the support and input of the farmers and other members of the local farming community in Exmoor through the face-to-face interviews and their participation so our thanks to them is substantial. The authors would like to thank the anonymous reviewers for their helpful and informative comments. All errors remain the responsibility of the authors.

REFERENCES

1. Baldock, D, J Dwyer, and JM Sumpsi-Vinas. 2002. Environmental integration and the CAP. Brussels: Commission of the European Communities.
2. Berkes, Fikret, J Colding, and C Folke. 2000. Rediscovery of traditional ecological knowledge as adaptive management. Ecological Applications 10 (5): 1251–1262.
3. Boatman, N, P Gaskell, C Short, H Parry, N Jones, S Bishop, J Mills, and J Dwyer. 2008. A Review of Environmental Benefits supplied by Agri-Environment Schemes, Report to the Land Use Policy Group by the Central Science Laboratory and Countryside and Community Research Institute. London: Natural England.
4. Carlsson Lars and Fikret Berkes. 2005. Co-management: concepts and methodological implications. Journal of Environmental Management 75: 65–76.
5. Carruthers, P, N Thomson, T Carroll, S Webster, A Harper, and I Soane. 2009. Developing the English Uplands: a report to the Commission for Rural Communities' Inquiry into the Future for England's Upland Communities. Cheltenham: CRC.
6. Commission for Rural Communities. 2008. An Inquiry into the Future for England's Upland Communities: Summary note. Cheltenham: CRC.
7. Commission for Rural Communities. 2010. The Future for England's Upland Communities – final report. Cheltenham: CRC.
8. Condliffe, Ian. 2009. Policy change in the uplands. In Drivers of Environmental Change in the Uplands, ed. Aletta Bonn, Tim Allott, Klaus Hubacek, Jon Stewart, 59–90. London: Routledge.

9. Defra, . 2011. June survey of agriculture and horticulture. http://www.defra.gov.uk/statistics/foodfarm/landuselivestock/junesurvey/. Accessed 22 October 2011
10. Defra Agricultural Change and Environment Observatory. 2009. Farm Practices Survey 2009 – Uplands And Other Less Favoured Areas (England) Attitudes To The Future. York: ACEO.
11. Dwyer, J. 2010. UK land use futures: policy influence and challenges up to 2060. Land Use Policy.
12. English, Nature. 1996. Upland Policy Review. Peterborough: English Nature.
13. English, Nature. 2002. SSSI condition survey reports: upland habitats. Peterborough: English Nature.
14. Exmoor National Park. 2007. Exmoor National Park Management Plan. Dulverton: Exmoor National Park Authority.
15. Eriksson, Camilla. 2011. What is traditional pastoral farming? The politics of heritage and 'real values' in Swedish summer farms (fäbodbruk). Pastoralism: Research, Policy and Practice 1: 25.
16. Franks, J, D Harvey, and C Scott. 2008. Farm Business Survey 2006–2007 Hill Farming in England. Newcastle: Newcastle University Rural Business Research.
17. Gaskell, , Peter, N Boatman, J Dwyer, J Mills, J Ingram, C Shor, and D Turley. 2008. Environmental observatory update – the implications of the 2003 CAP reform for the environment in England. A report to Defra's Agricultural Change and Environment Observatory (ACEO). Cheltenham: CCRI.
18. Jacobson, C, K Hughey, W Allen, S Rixecker, and R Carter. 2009. Toward more reflexive use of adaptive management. Society and Natural Resources 22: 484–495.
19. Kasperson, R. 2006. Rerouting the stakeholder express. Global Environmental Change 16: 320–322.
20. Lawton, JH, PNM Brotherton, VK Brown, C Elphick, AH Fitter, J Forshaw, RW Haddow, S Hilborne, RN Leafe, GM Mace, MP Southgate, WJ Sutherland, TE Tew, J Varley, and GR Wynne. 2010. Making space for nature: a review of England's wildlife sites and ecological network. Report to Defra.
21. Leys, AJ, and JK Vanclay. 2011. Social learning: a knowledge and capacity building approach for adaptive co-management of

contested landscapes. Land Use Policy 28: 574–584.

22. Likens, G. 1992. The ecosystem approach: its use and abuse. In Vol 3 Excellence in Ecology, ed. Kinne O, Germany Ecology Institute: Oldendor.
23. Lowe Philip and Neil Ward. 2007. Sustainable Rural Economies: Some Lessons from the English Experience. Sustainable Development 15: 307–317.
24. Millennium Ecosystem Assessment. 2005. Ecosystems and Human Well-being: Synthesis. Washington, DC: Island Press.
25. National Ecosystem Assessment. 2011. UK National Ecosystem Assessment. Cambridge: UNEP-WCMC. http://uknea.unep-wcmc.org. Accessed 22 October 2011
26. England, N. 2009. A Vision for the Uplands. Peterborough: Natural England.
27. Natural England. 2010. South West uplands. http://www.naturalengland.org.uk/regions/south_west/ourwork/southwestuplands/default.aspx. Accessed 22 October 2011
28. Natural England. 2011. SSSI condition summary. http://www.sssi.naturalengland.org.uk/Special/sssi/report.cfm?category=N. Accessed 10 November 2011
29. Nori, Silvia, and Gemini Michele. 2011. The Common Agricultural Policy vis-à-vis European pastoralists: principles and practices. Pastoralism: Research, Policy and Practice 1: 27.
30. Olsson, Per, Folke Carl, and Berkes Fikret. 2004. Adaptive co-management for building resilience in socio-ecological systems. Environmental Management 34 (1): 75–90.
31. Reed, M. 2008. Stakeholder participation for environmental management: A literature review. Biological Conservation 141: 2417–2431.
32. Reed, M, A Bonn, W Slee, N Beharry-Borg, J Birch, I Brown, TP Burt, D Chapman, PJ Chapman, GD Clay, SJ Cornell, EDG Fraser, JH Glass, J Holden, JA Hodgson, K Hubacel, B Irvine, N Jin, MJ Kirkby, WE Kunin, O Moore, D Moseley, C Prell, MF Price, CH Quinn, S Redpath, C Reid, S Stagl, LC Stringer, M Termansen, S Thorp, W Towers, and F Worrall. 2009. The future of the uplands. Land Use Policy 26: 204–216.

33. Ribot, , and Jesse. 2006. Choose democracy: environmentalists' socio-political responsibility. Global Environmental Change 16: 115–119.

34. Simoncini, R, G Borrini-Feyerabend, and B Lassen. 2008. Policy guidelines on governance and ecosystem management for biodiversity conservation, report of the Governance and Ecosystem Management for the Conservation of Biology (GEMCONBIO) project, Final report. www.gemconbio.eu . Accessed 10 October 2011

35. Short, , and Christopher. 2008. The traditional commons of England and Wales in the twenty-first century: meeting new and old challenges. International Journal of the Commons - European Commons Special Issue. 192–221.

36. Short, C, R Griffiths, and J Phelps. 2010. Inspiring and Enabling Local Communities: an integrated delivery model for Localism and the Environment. CCRI: Report to Farming and Wildlife Advisory Group and Natural England. Cheltenham.

37. Swales, V, R Eaton, V Castan-Broto, R Deane, B Warmington, S Parker, and M Rayment. 2004. An assessment of the impacts of hill farming in England on the economic, environmental and social sustainability of the uplands and more widely. Defra: Report to Defra. London.

38. The World Bank. 1999. Report from the International Workshop on Community-Based Natural Resource Management (CBNRM). Washington, DC. 10–14 may 1998

39. Turner, M, K Robbins, and P Silcock. 2008. Hill Farming Systems in South West England: economic viability and the delivery of pubic goods. Exeter: University of Exeter.

Chapter 5

Invasive Species in Europe: Ecology, Status, and Policy

Reuben P Keller[1], Juergen Geist[2], Jonathan M Jeschke[3], and Ingolf Kühn[4]

[1]Program on the Global Environment, University of Chicago, Chicago IL 60637, USA

[2]Aquatic Systems Biology Unit, Department of Ecology and Ecosystem Management, Technische Universität München, Mühlenweg 22, 85354 Freising, Germany

[3]Department of Biology II, Ludwig-Maximilians-University Munich, Grosshaderner Str. 2, 82152 Planegg-Martinsried, Germany

[4]UFZ, Helmholtz Centre for Environmental Research--UFZ, Department of Community Ecology, Theodor-Lieser-Str. 4, 06120 Halle, Germany

ABSTRACT

Globalization of trade and travel has facilitated the spread of non-native species across the earth. A proportion of these species become established and cause serious environmental, economic, and human health impacts. These species are referred to as *invasive*, and are now recognized as one of the major drivers of biodiversity change across the globe. As a long-time centre for trade, Europe has seen the introduction and subsequent establishment of at least several thousand non-native species. These range in taxonomy from viruses and bacteria to fungi, plants, and animals. Although invasive species cause major negative impacts across all regions of Europe, they also offer scientists the opportunity to develop and test theory about how species enter and leave communities, how non-native and native species interact with each other, and how different types of species affect ecosystem functions. For these reasons, there has been recent growth in the field of invasion biology as scientists work to understand the process of invasion, the changes that invasive species cause to their recipient ecosystems, and the ways that the problems of invasive species can be reduced. This review covers the process and drivers of species invasions in Europe, the socio-economic factors that make some regions particularly strongly invaded, and the ecological factors that make some species particularly invasive. We describe the impacts of invasive species in Europe; the difficulties involved in reducing these impacts, and explain the policy options currently being considered. We outline the reasons that invasive species create unique policy challenges, and suggest some rules of thumb for designing and implementing management programs. If new management programs are not enacted in Europe, it is inevitable that more invasive species will arrive, and that the total economic, environmental, and human health impacts from these species will continue to grow.

INTRODUCTION

Globalization has integrated widely dispersed human communities into a worldwide economy. This process provides many benefits through the movement of people and goods, but also leads to the intentional and unintentional transfer of organisms among ecosystems that were previously separate [1]. Some of these species become established beyond their native range, a subset of these spread, and some of these have negative impacts and are termed invasive [2]. Although not all species moved beyond their native range become established [3], the large number of species transported and the range of pathways that move species mean that non-native species are now recognized as one of the major drivers of global biodiversity loss. They also cause significant damage to economies and human health [4-7].

Europe has been a centre for international trade for many centuries and has consequently seen the establishment of a large number of species. Some of these species have positive effects, including a subset of those introduced to enhance fisheries. Many other species, however, cause large negative impacts. These species cover a broad taxonomic range - from viruses and bacteria to fungi, plants, and animals - and affect all European nations and regions [6]. Where timelines are available, the number of non-native species established in Europe is generally increasing exponentially in freshwater [8] and terrestrial ecosystems [5, 8-11]. This pattern is consistent with exponential increases in trade and travel [5]. Without increased efforts to manage pathways of introduction, the number of invasive species will continue to grow. Indeed, because there is often a significant lag period between species introduction and spread, it is likely that many future invasions have already been set in motion [12,13]. Consequently, the task of designing policies to reduce the transport and release of non-native species, and to manage those already established, has become a large priority for both national governments within Europe, and for the European Union [14,15].

In contrast to their negative impacts, non-native species offer large opportunities for ecologists to test fundamental theory [16,17]. In many cases, established non-native species spread rapidly into new ecosystems. These species interact with native species through competition, predation, herbivory, introduction of disease, and resource use. When carefully studied, the effects of these perturbations allow insight into basic ecological processes. For example, community ecology seeks to understand why certain species communities exist (or do not), and how species within those communities interact. Because invasions involve new species entering communities and interacting with resident species, they offer insight into these patterns and processes. The very real need to better understand the process of invasion so that native habitats can be conserved, combined with opportunities to test fundamental ecological processes, has led to huge recent growth in the field of invasion biology (e.g.[18]), including in Europe.

The goal of this paper is to review the ecology and current status of non-native species in Europe, and to outline a framework for policy to reduce their number and impacts. We begin with a description of the *invasion process* through which non-native species are introduced, become established, and may then go on to have negative impacts. This structures the sections that follow which cover the pathways of species introduction, the characteristics of habitats and species that lead to establishment, the number of non-native species in Europe, and their impacts. Finally, we describe some characteristics of species invasions that create unique challenges compared to other environmental issues. We propose some general principles to govern the policy response to the risks from invasive species. Throughout the paper, we discriminate among three groups of organisms: terrestrial animals, terrestrial plants, and aquatic organisms. This distinction corresponds to the published literature, which in many cases is quite divergent among the groups. Our sections about aquatic organisms include both marine and freshwater species. We primarily review material specific to species invasions in Europe, but include work conducted elsewhere as necessary.

THE INVASION PROCESS

To become invasive, a species must pass through a number of transitions (e.g. [2, 19, and 20]). First, it must be entrained in a *pathway* and survive transit, where a pathway is a human-mediated process that facilitates the movement of species from one region to another. When it arrives in a region beyond its native range due to direct or indirect human intervention, it is referred to as *introduced*. Species that are not able to maintain self-sustaining populations, but that are occasionally found beyond cultivation, are often termed *casual* species. Next, if a species survives, escapes, and begins reproducing without direct human intervention it is referred to as *established*. Finally, we refer to a species as *invasive* if it spreads widely and causes measurable environmental, economic, or human health impacts.

Although the process described above is common to all invasions, different terminologies have been proposed, and these are often associated with different taxa or regions. For example, botanists may speak of an *invasive* plant as one that exceeds some predetermined rates of spread, whether or not it causes negative impacts (e.g. [19]). While we acknowledge the range of terminologies available, throughout this paper we use the definitions given in the previous paragraph.

The proportion of introduced species that become established can be quite low, as can the proportion of established species that spread and become invasive. These proportions vary with the taxonomy of the species in question, and the regions to which they are introduced. The 'tens rule' was proposed by Williamson [3] as a rule of thumb to approximate the proportion of species that make it through each step in the invasion process. This rule holds that approximately 10% of introduced species will become established, and approximately 10% of those species will become invasive. Hence, if 100 alien species were introduced, the tens rule holds that one will become invasive. The tens rule was developed with a focus on terrestrial plants, but has been applied by other authors to a wide range of species, often without thoroughly evaluating its validity. For example, recent studies show that for many animal taxa, the percentage of introduced species that become established and the percentage of established species that become invasive can exceed 50% [5, 21]. According to the definition of invasive given above, Europe now contains >100 terrestrial vertebrates, >600 terrestrial invertebrates, >300 terrestrial plants, and >300 aquatic species that have become invasive [7].

The following sections are organized to follow the invasion process. In each, we describe the current state of knowledge about non-native terrestrial animals, plants, and aquatic organisms in Europe. We discuss the dominant pathways of introduction and the characteristics of ecosystems and species that most often lead to establishment. Next, we give estimates of the number of established non-native species from each group in Europe, and describe their impacts.

INTRODUCTION PATHWAYS

There is a long history of classifying the pathways of non-native species. One of the oldest classifications was developed by Thellung [22, 23] for Central Europe. More recent schemes are from [9] and [24], and reviews and syntheses are from [25] and [11]. Hulme et al. [9], the most recent classification from a European perspective, identified three general mechanisms through which non-native species may enter a new region: importation as or with a commodity, arrival with a transport vector, and dispersal by the species themselves, either along infrastructure corridors (e.g. roads, canals) or unaided. The first of these, transportation as or with a commodity, arises from direct human movement of goods. Transportation *as* a commodity occurs when people identify a species as having desirable qualities and intentionally move it beyond its native range. Species introduced as commodities can be released intentionally (e.g. stocking animals to create populations for fishing and hunting, introduction of species as biological control agents) or can escape unintentionally (e.g. ornamental plants reproduce and spread beyond gardens, fish in aquaculture facilities escape). Species transportation *with* a commodity occurs when traded organisms arrive contaminated by non-native species, including diseases and parasites. These contaminants are not introduced intentionally, but may escape to become established and invasive. For example, the crayfish plague disease (*Aphamomyces astaci*) was introduced to Europe on North American crayfish imported for aquaculture. This disease has escaped, become established, and now infests and endangers native crayfish populations across Europe [26]. Another example is the Asian tiger mosquito (*Aedes albopictus*), which is native to South-east Asia but has spread to at least 28 countries, including several European countries, on ships as a contaminant of the trade in used car tires [27].

The second mechanism, arrival with a transport vector, refers to species that hitchhike on human modes of transport to reach regions beyond their native range. Modes of transport include ships (e.g. species in ballast water or attached to hulls), airplanes (e.g. contaminants of cargo, diseases/pests on food carried by passengers), and automobiles (e.g. as seeds caught in mud on tires). The influence of ships has been particularly strong in aquatic ecosystems, with many marine and freshwater species having become harmful invaders after dispersal in or on ships [28]. In Europe, this has included the spread of zebra mussel (*Dreissena polymorpha*) from the Ponto-Caspian Basin to the Atlantic and U.K. through the river and canal network.

The third mechanism is dispersal by the species themselves, either along infrastructure corridors or unaided. Corridor dispersal occurs when organisms move along canals, railways, roads, and other linear habitats created by humans. Examples include the introduction of species from the Red Sea into the Mediterranean Sea through the Suez Canal [29]. Unaided dispersal occurs when a non-native species becomes established in a neighbouring or nearby ecosystem, and then spreads without human intervention. An example is the continuing spread of invasive horse-chestnut leafminer moth (*Camariella ohridella*)

[30] across Europe. Although this species was only introduced into a limited area, it has developed large populations and spread widely. Table 1 lists a number of important pathways that have transported species that are now established to Europe.

Table 1: Some important pathways of introduction for non-native terrestrial animals, terrestrial plants, and aquatic organisms

Group	Sub-group	Dominant pathways	References
Terrestrial vertebrates	Mammals	Intentional introduction as commodity (for hunting, 'fauna improvement', fur farming, as pets, or for zoos), then either intentional release or accidental escape	[9,31]
	Birds	Intentional introduction as commodity (for hunting, 'fauna improvement', as pets, or for zoos or bird parks), then either intentional release or accidental escape	[9,32]

	Reptiles/ amphibians	Intentional introduction as commodity (for 'fauna improvement', as pets, food source, or biological control agents), then either intentional release or accidental escape	[9,32]
Terrestrial invertebrates	Insects	Unintentional introduction as contaminants or stowaways, sometimes deliberate release as biological control agents	[9,33]
	Other	Unintentional introduction as contaminants or stowaways	[33]
Terrestrial plants	Vascular plants, mosses, and lichens	Intentional introduction as commodity for garden trade (ornamentals), horticulture, unintentional introduction as contaminant of plants introduced for agriculture and ornamental trade (e.g., soil contaminants in plant pots)	[9,10,34,126]

Aquatic organisms	Fishes	Intentional introduction for aquaculture, stocking to improve recreational and commercial fisheries (including illegal stocking), as well as for weed and mosquito control, unintentional introductions with ship ballast water, ornamental species, fishing bait releases	[75,127]
	Crustaceans	Intentional introduction for aquaculture, ornamental reasons (Decapoda), unintentional introductions with ship ballast water, canals	[38,128-130]
	Mollusks	Unintentional introductions with shipping, waterways, accidental (e.g. during fish stocking), but also from garden pond and aquarium trade	[131,132]

	Plants	Intentional introduction for ornamental (aquarium and watergarden) trade, often further spread by boats and waterbirds	[133,134]

Pathways of introduction for organisms established in Europe. Pathway lists given are not comprehensive and were chosen to give an indication of the total range of vectors, not necessarily those that are most important for each group.

Keller *et al. Environmental Sciences Europe* 2011 23:23, doi: 10.1186/2190-4715-23-23

Terrestrial Animals

Most terrestrial vertebrate animals established in Europe (or elsewhere throughout the world) were intentionally introduced as commodities, e.g. by the pet trade, the live food trade, or as stock for the trade in fur pelts [9, 31-33]. Although some of these pathways have been modified and restricted to reduce the risk of invasion, many remain very active. For example, the pet trade remains a dominant pathway for the introduction of new invasive species to Europe [31, 32].

Many other terrestrial animal species, especially invertebrates, have been introduced across Europe unintentionally, mostly as stowaways or contaminants on traded products, on vehicles (e.g. ships), or as diseases/parasites of plants, animals, and humans [9, 31-33]. In general, much less is known about introductions of species that were introduced unintentionally because they are usually not recorded until they become established.

Terrestrial Plants

Almost two thirds (62.8%) of the established plant species in Europe were introduced intentionally for ornamental, horticultural, or agricultural purposes. The remaining species were introduced unintentionally, mostly associated with transport vectors, or as contaminants of seeds and other commodities [34]. Of the terrestrial plant species that have escaped from human cultivation, some were intentionally released (i.e. planted in the wild to 'improve' the landscape), some were contaminants or stowaways, and only a few arrived unaided [9]. Consistent with increases in international trade, there has been a steady increase in the number of established non-native terrestrial plant species discovered in Europe, especially since 1800. Currently, an average of 6.2 species not native to any part of Europe is newly recorded as established each year. An average of 5.3 European species are found in parts of the continent outside their native range each year [10].

Much less is known about the introduction and spread of non-native lower plants and fungi, and about changes in the number of non-native species in Europe over time. It is known that these taxa can have enormous impacts, with perhaps the most damaging examples being diseases of crops and livestock. The potato famines that occurred across Western Europe during the nineteenth century, for example, were caused by the invasive potato blight (*Phytophthora infestans*) [35], introduced from North America.

Aquatic Organisms

The pathways of introduction for aquatic organisms are generally known with less precision than those for terrestrial organisms. This arises partly because many aquatic species are introduced unintentionally with few (if any) records kept. Additionally, the difficulty of sampling in marine and freshwater environments means that a species may be well established, and may have spread from its initial site of introduction, before it is recorded.

Shipping has been by far the dominant pathway for the introduction of non-native marine species to the European Atlantic coast (47% of established non-native species) and to the Baltic Sea (45%,[36]). This pathway has also been a significant factor in freshwater animal introductions to Europe (25% of established non-native species, [37]). The shipping network creates connections among aquatic ecosystems across the globe, and organisms are frequently transported in the ballast water of ships, or attached to hulls as fouling organisms [28]. Ballast water is taken on to increase a vessel's weight when it is not fully laden with cargo. As this water is taken on, any organisms in the water are also sucked in. Vessels then travel to subsequent ports, and surviving organisms can be discharged with ballast water if the vessel takes on more cargo.

The opening of canals that link previously isolated water bodies has created many opportunities for the introduction and spread of non-native species. In the Mediterranean Sea, 54% of established non-native species arrived by dispersing through the Suez Canal [36]. Canals have also had a profound impact on the establishment and spread of non-native freshwater species in Europe, and this impact is tightly linked to shipping. There are now river and canal connections running from the Black Sea across Europe to the mouth of the Rhine River, and north to the Baltic Sea [38]. These connections have served as invasion corridors for many species native to the Ponto-Caspian into Western and Northern Europe. It is estimated that 8% of non-native freshwater animal species in Europe arrived by using natural dispersal mechanisms to move through canals [37]. In addition, many of the species that arrived through shipping (see previous paragraph) could only do so because of the existence of canals.

Despite the influence of shipping and canals, the most important pathways for the introduction of non-native freshwater animal species to Europe have been stocking (30% of species) and aquaculture (27%) [37]. Stocking has been largely of fish to create new wild populations, while aquaculture introductions have arisen from the unintended escape of farmed species and their associated organisms (e.g. parasites). Aquaculture has also been important for introductions of marine species to the Atlantic coast, Baltic Sea, and Mediterranean, accounting for 24%, 18%, and 11% of established species, respectively [36].

The final pathways mentioned here are the trades in ornamental (predominantly aquarium and watergarden) and aquaculture species. Ornamental introductions have been especially important in freshwater ecosystems, accounting for 8% of established non-native animal species. Ornamental introductions also appear to be by far the dominant pathway for introduction of aquatic plants. For example, in Great Britain 22 of 31 established non-native freshwater plant species were introduced for the ornamental trade [8]. The aquaculture trade has unintentionally introduced a large number of non-native aquatic species as contaminants of intentionally introduced species such as fish or shellfish. This is true for both marine and freshwater habitats. For example, the unintentional introduction and spread of the brown algae *Sargassum muticum*, the Japanease kelp *Undaria pinnatifida*, and the snail *Ocinebrellus inornatus*, as well as the oyster parasites *Mytilicola orientalis*and *Myicola ostreae*, all occurred because these species inadvertently arrived associated with marine shellfish imported from Asia to Europe for aquaculture [36].

CHARACTERISTICS OF HIGHLY INVADED REGIONS

The number of invasive species found in a region depends on the number of species that have been introduced, the proportion of introduced species that have become established, and the proportion of established species that have gone on to cause impacts. When investigating differences among regions, invasion biologists have generally left aside the pathways and process of introduction and focused instead on the proportions of introduced species that become established, and of established species that become invasive [39]. Ecosystems where these proportions are high have been assumed to be highly invasible, while others have been deemed relatively resistant.

Different theories have been proposed to explain why some regions appear more invasible than others. Perhaps the most influential has been the *biotic resistance* hypothesis, an early champion of which was Charles Elton, often referred to as the scientist who founded the field of invasion biology (e.g. [19]; although we note that Charles Darwin [40] and other biologists had already written about the spread and impacts of non-native species, [11,41]). This hypothesis holds that regions with high biodiversity and a relatively low level of disturbance, especially disturbance from humans, are more resistant to establishment by non-native species [42]. The rationale is that less diverse and/or more disturbed ecosystems are likely to have more vacant niches that introduced species can inhabit. Although this is an intuitively appealing argument, there has been little empirical evidence generated to support it. In fact, especially at larger spatial scales, there is increasing evidence that highly diverse habitats are actually more prone to non-native species establishment (e.g. [43-45]. Several authors have attempted to reconcile the contrasting theory and evidence, but no consensus has yet been reached (e.g. [46,47]). A large obstacle to finding this consensus comes from the difficulty of quantitatively assessing levels of disturbance and the presence of vacant niches.

As well as trying to reconcile theory and observed patterns in species establishment, ecologists are now paying more attention to the introduction process. Recent results show that the number of species introduced to a region may be at least as important as the invasibility of the region in determining how many species become established [48-52]. This is discussed further in the following sections.

Terrestrial Animals

There is a particular lack of support for the biotic resistance hypothesis when terrestrial animals are considered (see [52] and references therein). It has become clear in recent years that the key difference among regions with different numbers of established terrestrial animals is the number of species that have been introduced. But which regions are the ones that receive more introductions than others? The answer is that regions with high human impact typically receive more species introductions than other regions, and that this leads to them containing more established species. For example, 12 non-native mammal species have established in France, nine in Germany, but just two in Portugal. Contrary to what would be predicted from the biotic resistance hypothesis, it is not easier for introduced mammals to establish in countries with high human impact, but these countries host more non-native mammals than other countries because they have received more species introductions [52]. Differences among the number of established birds in European countries can also be best explained by differences in number of introduced species [51].

Plants

Patterns of plant invasion in Europe offer little support for the biotic resistance hypothesis (e.g.[44,53,54]). Instead, apart from broad habitat type, the number of species introduced and their propagule pressure appear to be the most important determinants of the number of established non-native species in any given region (e.g. [55,56]). The most invaded habitats in Europe are in heavily transformed landscapes such as agricultural land, coniferous forests, urban areas, and dump or construction sites [57]. In contrast, natural and semi-natural environments such as broad leaved and mixed forests, pastures, natural grasslands, moors, heathlands and peatbogs have remained relatively uninvaded [57]. This pattern is consistent with that observed for terrestrial animals - that sites experiencing high levels of human disturbance and high propagule pressure tend to be the most invaded. Disturbance increases plant invasions because it leads to loss of native species that could compete with introduced non-native species, and because it increases availability of resources [58]. High propagule pressure occurs in the same regions because human activities lead to many plant introductions [59].

The highest proportions of established terrestrial plant species in Europe occur in agricultural landscapes, especially in eastern Britain, northern France, Central and Eastern Europe, and the Po floodplain in Italy. In contrast to a global pattern of Mediterranean-type ecosystems being highly invaded, the European Mediterranean biogeographic region is relatively uninvaded, probably due to a long history of human presence and prehistoric introductions in the Mediterranean Basin, which may make its ecosystems relatively resistant against recently introduced species [57]. Additionally, the Mediterranean Basin acted as more of a donor than recipient region for species introductions during the colonization of the New World [60].

It has been argued that harsh environments (e.g. alpine habitats) might not be suitable for non-native species. However, these are also often the habitats that experience low propagule pressure [61]. Hence, it is clear that the intensity of human activities that increase or facilitate propagule pressure, pathways of introduction, intensity of disturbance, and eutrophication, are important determinants of non-native plant invasions. For many taxa in Europe, it is even more important than climate or other features of the physical environment [62].

Aquatic Organisms

European aquatic ecosystems containing the highest numbers of non-native species tend to be those with high connectivity to other ecosystems, high frequency of human access (e.g. for transportation or recreation), and high disturbance. These include boat harbours, recreational areas at lakes (jetties etc.), and the many canals that now cross Europe. More remote water bodies, including mountain lakes and headwater streams, tend to be least and last invaded. Thus, propagule pressure can largely explain the intensity and diversity of established non-native species in aquatic environments [3, 20]. In marine ecosystems, the number and frequency of pathways, tidal movements, availability of empty niches, and availability of different substrate types for settlement are the main factors that determine susceptibility to invasion, with highest rates of non-native species establishment typically found in shallow coastal zones [36]. Consequently, marine ecosystems with high numbers of established species in Europe include the eastern Mediterranean with hundreds of introductions through the Suez Canal [63], as well as the Gulf of Finland, the Gulf of Riga, the coastal lagoons [64-66], and the Oosterschelde estuary [67]. Of the 737 non-native multicellular animal species recorded from European seas, 569 have been found in the Mediterranean, 200 along the Atlantic coast (Norway to the Azores, including the UK and Ireland), and 62 in the Baltic Sea [36]. Numbers in the Mediterranean are highest because of the Suez Canal, the role of the Mediterranean as a long-time hub of international shipping, and a surge in development of mariculture [36].

CHARACTERISTICS OF INVASIVE SPECIES

An alternative perspective comes from asking whether there are traits of non-native species that are associated with successful passage through the invasion process. Ecologists have been asking this question for several decades (e.g. [68]), often concentrating on intuitive life-history traits such as early reproduction, high reproductive output, or a non-specialized diet (reviewed by [69,70]). This work has recently become more important because many nations, including several in Europe and the European Union, have begun to develop risk assessment programs for non-native species [71-75]. The development of risk assessment tools begins with the search for patterns in species traits that are associated with successful passage through the invasion process. If robust patterns are identified, they can be applied to non-native species to determine the likelihood that they will become established, spread, and/or become invasive (see [2, 76, and 77] for reviews). In this way, knowledge of non-native species traits can support pro-active efforts to prevent new invasions.

Although the search for traits of invasive species has been fruitful, recent results have shown that propagule pressure can be confounded with invasiveness [39]. As for the discussion above that focused on differences in invasibility among regions, invasion biologists have traditionally left aside introduction and focused on establishment and spread when looking for differences between the traits of invasive and non-invasive species. Recent studies have challenged this by showing that those species most likely to establish are often those introduced in the highest numbers and most frequently. This does not mean that the search for traits of invasive species is not worthwhile, but does indicate that additional factors are important.

Terrestrial Animals

Recent studies of the characteristics of invasive terrestrial animal species have shown that invasive species tend to have been introduced in higher numbers and more frequently than non-invasive species [69, 78]. Mammals and birds that are hunted by humans are more frequently invasive than other species of mammals and birds because they have been more frequently introduced than other species, even though their establishment success is not higher than that of other species [69]. The same is true for mammals and birds with large native ranges which also become invasive more frequently than species with smaller native ranges. Their establishment success has not been shown to be higher than that of non-native species with smaller native ranges, meaning that this pattern is best explained by their increased frequency of introduction [69, 79].

Terrestrial animals living in association with humans tend to become invasive more often than other species [69]. Good examples are the Norway rat (*Rattus norvegicus*) which reaches extremely high population densities in cities; the rose-ringed parakeet (*Psittacula krameri*), native to Africa and Asia and also often very abundant in human settlements; and the harlequin ladybird (*Harmonia axyridis*), native to Asia and infamous for its large aggregations in buildings during winter [6]. Thus, a clear understanding of human activities, both in terms of propagule pressure and the location of human settlements, is very important for understanding patterns of establishment, spread, and harm for non-native terrestrial animals.

There are also species-level biological traits linked to terrestrial animal invasiveness. For example, behavioural flexibility as expressed by brain size was among the best predictors of invasiveness in a study of non-native birds [80]. Mammals and birds with high ecological flexibility (indicated by the number of different types of food they consume or the number of different types of habitat they use) also tend to be more invasive than other species [69, 81, and 82]. Thus, species that are relatively more behaviourally and ecologically flexible tend to become invasive more often than other species.

Terrestrial Plants

Several factors are related to the invasion success of individual plant species. First, non-native plants that have been introduced and/or planted more frequently (i.e. higher propagule pressure) are more likely to become established and to have a larger range (e.g. [83]). Second, residence time (i.e. time since introduction) in the non-native range is important, with those species that have been present for longer tending to have larger ranges [84]. This is an effect of having the opportunity to fulfil more life cycles and also simply having the time to spread further. The importance of residence time is also associated with propagule pressure, as species that were introduced a long time ago are likely to have been introduced many times since the first introduction. Third, species with larger native ranges are more likely to successfully establish beyond their native range. Similar to terrestrial animals, this is presumably associated with a higher probability that the species will be accidentally transported [85]. Additionally, species with a large native range are more likely to have a strong climate match to at least part of Europe, making them pre-adapted to survive there. Fourth, once terrestrial plant species have been introduced to the new range, traits of the species are important for determining whether they will successfully establish, spread, and cause harm. Traits known to promote passage through the invasion sequence include long flowering season, being an annual, vegetative spread, having multiple dispersal vectors [85], high maximum relative growth rate, and high resource allocation to shoots and leaves [86,87].

Many studies that have attempted to relate biological traits to invasiveness have explained little of the variation and have neglected trait interactions. Including interactions among traits (i.e. explicitly considering that one trait value might have a different influence on invasion success in the presence of other traits) can result in much better explanatory models. Küster et al. [88] found that trait interactions accounted for >40% of the variation that could explain invasion success of non-native terrestrial plants in Germany. Interestingly, long flowering season was beneficial for self-pollinated species, but was disadvantageous for wind pollinated species, and had no effect on insect-pollinated species. Furthermore, the effect of timing of the end of the flowering season on invasion success differed among plant species with different vegetative reproduction strategies or different levels of ploidy (number of chromosome sets in the cell). Thompson and Davis [89], however, argue that such analyses tell us very little because successful invaders do not differ in their traits from those of widespread native plant species. Despite this critique, incorporating statistical interactions among traits should increase our knowledge of characteristics that make a species likely to expand or contract its range, whether non-native or native.

Aquatic Organisms

As for terrestrial animals and plants, there are some general rules that separate non-native aquatic species that successfully pass through the invasion sequence from those that do not. Some of the characteristics that influence invasion success are associated with biological traits whereas others are closely linked to interaction with humans. For example, species introduced intentionally (and hitchhiking species associated with them) because they have desirable attributes tend to be more successful than undesired species. Prominent examples in marine environments include the introductions of alien shellfish species (e.g. *Crassostrea gigas* introduced to France from Japan) for mariculture, which have arrived with several associated parasites and algae. Additionally, many of the most widespread non-native aquatic species in Europe are generalists that can tolerate a wide range of environmental conditions such as water temperature and salinity. European brackish water systems hold a great diversity of invaders which may be due to their poor native species richness [90] and the great ecological plasticity of the non-native species that have established. In addition to the breadth of ecological niches, similarity of environmental conditions in the donor and the receiving region can also be crucial [28]. For example, most of the 569 non-native species in the Mediterranean are thermophilic and originated from tropical waters in the Indo-Pacific, the Indian Ocean, the Red Sea, and pan-tropical regions [36].

Differences in life history and reproduction can differentiate between invaders and rare species. This is evident in freshwater unionid mussels, which are among the most critically imperiled freshwater taxa both in North America and Europe [91, 92]. These species produce glochidia larvae that need to attach to a suitable fish host to survive. The high degree of specialization and the complex life cycle of unionids probably contribute to the decline in this group. In contrast, invasive mussels of the genus *Dreissena* are less specialized and produce free veliger larvae, allowing for a higher rate of dispersal through passive transport (e.g. in the ballast water of ships). Non-native marine species, which have been predominantly introduced to Europe through shipping, are also more likely to have larval stages that are tolerant of conditions in ships.

Reproduction rates tend to be higher in invasive aquatic species compared to those in most non-invasive species (e.g. [93]). High reproduction can facilitate rapid spread and secondary introductions into other areas. Mode of feeding can also be important, with filter-feeding freshwater macroinvertebrates in Europe and North America known to be more successful at invading than predator macroinvertebrates [94]. This has the impact of enhancing energy flow between benthic (i.e. bottom) and pelagic (i.e. open water) regions because algae that are produced mostly in the pelagic zone are consumed by the benthic filter feeders. For intentional introductions of fishes, where large predator species tend to be most popular, competition and top-down regulation may be more important. Overall, effects of non-native species tend to be greater when they establish in high abundance and have strong functional distinctiveness from native species [95].

NUMBER OF ESTABLISHED NON-NATIVE SPECIES IN EUROPE

At least several thousand non-native species are now established in Europe [6]. These include species not native to any part of Europe, as well as species native to one part but now established in another. The following sections give estimates of the number of species in different habitat categories that are established in Europe and not native to any part (unless stated otherwise). These figures should be seen as low estimates of the true numbers of established species because only recorded species are included; it is likely that many additional species are established but not yet recorded.

Terrestrial Animals

According to the DAISIE database, there are 33 non-native established mammal species [31] and 77 established bird species in Europe [32]. These figures are probably quite accurate because these taxa are relatively large and easy to distinguish from native species. For the same reason, the estimate of 55 established reptiles and amphibians in Europe [32] is also probably quite accurate. In contrast, estimates for invertebrates are likely to be more severe underestimates because these species are more difficult to collect and identify. Within terrestrial invertebrates, data for insects tend to be more accurate than those for other invertebrates [33]. Insects are also the dominant group among non-native terrestrial invertebrates in Europe: of 1,522 established species, 1,306 (86%) are insects [33]. This high proportion is not unexpected, however, as 85% of the world's known invertebrates are insects [96].

Terrestrial Plants

Terrestrial plants are generally well sampled, but it can be difficult to assess total numbers of established non-native species because the same species is often given different scientific names in different parts of Europe. According to the DAISIE database [10], 5,789 plant species have been recorded from the wild (not necessarily established) in at least one European country to which they are not native. These species come from 213 families and 1,567 genera, and include 2,843 species not native to any European country. A total of 3,749 plant species are known to be established in at least one European country to which they are not native, and 1,780 of these species are entirely non-native to Europe. We note that the numbers just given include all species recognized as non-native, irrespective of their date of introduction [97]. Traditionally, in those countries where records are available, botanists distinguish between species introduced before the European discovery of the Americas (1492) and those introduced later.

Aquatic Organisms

It is estimated that 737 non-native multicellular animal species from marine environments and 262 non-native freshwater animal species are now established in Europe [36, 37]. These comprise a wide range of taxa, including fishes, arthropods, mollusks, platyhelminthes, and annelids. For aquatic plants, it is estimated that at least 260 species not native to any part of Europe are established in inland waterways [34].

The number and diversity of non-native species is variable across different regions of Europe. For example, in Great Britain the 134 established non-native species in freshwater ecosystems are dominated by plants (31), fishes (18), non-decapod crustaceans (17), platyhelminths (15), and amphibians (11; full list in [8]). In Italy, the patterns are somewhat different, with the 112 non-native species from inland aquatic systems being dominated by fishes (38), non-decapod crustaceans (28), and gastropods (7; full list in [98]). In each case, it is reasonable to expect that non-native species from groups such as fishes and crayfishes are better represented in the data than records of species that are smaller and less often sampled (e.g. annelids).

Just seven non-native vascular plants have been identified in European marine ecosystems [34]. In contrast, numbers of non-native marine species (i.e. including animals and other multicellular organisms) are much larger. The three main marine biogeographic regions of Europe are the Mediterranean, the Atlantic coast, and the Baltic Sea; these contain 569, 200, and 62 established non-native species, respectively [36]. These species cover a large taxonomic range, from fishes to barnacles to plants.

IMPACTS OF NON-NATIVE SPECIES IN EUROPE

Invasive species have a large and diverse range of impacts in Europe. This diversity of impacts is mainly driven by the diversity of species, and makes generalized statements about types of impact difficult. However, it is clear that invasive species have significant negative impacts on many native species and almost all ecosystems, on the European economy, and on human health (recently reviewed by [7]). Economic impacts alone are estimated to be at least 12.5 billion EUR per year, and are probably over 20 billion EUR [99].

Terrestrial Animals

Ecological impacts of invasive terrestrial animal species include predation/herbivory, competition, transmission of diseases, and hybridization with native species. Economic effects include impacts on human infrastructure, human health, human social life, livestock, plant production, and forestry [7,100,101]. For example, Norway rats (*Rattus norvegicus*) predate many native species and have caused declines in native bird species and small mammal species. They are also a reservoir and vector of many diseases, including hepatitis E, leptospirosis, hantavirus, and Q fever. The invasive American mink (*Neovison vison*) is a competitor of the European mink (*Mustela lutreola*) which is now listed as endangered on the IUCN Red List of Threatened Species. Another example of an invasive competitor of a native species is the North American grey squirrel (*Sciurus carolinensis*, Figure 1A) that threatens the native red squirrel (*Sciurus vulgaris*), especially in the U.K. and Italy. The Canada goose (*Branta canadensis*, Figure 1B) is also an abundant invader. It hybridizes and competes with native geese, and its droppings can cause human health hazards and algal blooms. The Asian tiger mosquito (*Aedes albopictus*) competes with native mosquito species; its bites are a nuisance to humans, and it is a vector for diseases such as West Nile virus. Another invasive terrestrial invertebrate with severe impacts is the harlequin ladybird (*Harmonia axyridis*). Its tendency to overwinter in large aggregations inside buildings is a nuisance to many people, and the unpleasant odour of its body fluids can destroy the taste of wine. It also threatens native ladybirds and other European insect species [6]. Overall, invasive terrestrial invertebrate species cause costs of at least 1.5 billion EUR per year in Europe, and invasive terrestrial vertebrates cause costs of at least 4.8 billion EUR per year [99].

Figure 1: Examples of high-impact invasive species in Europe. (A) grey squirrel (*Sciurus carolinensis*), native to North America, © Jeschke; (B) Canada goose (*Branta canadensis*), also native to North America, © Jeschke; (C) giant hogweed (*Heracleum mantegazzianum*), native to the Caucasus region, © Denholm, NJ Dept of Agriculture, Bugwood.org; (D) round goby (*Neogobius melanostomus*), native to the Caspian, Black, and Azov Seas, © Aquatic Systems Biology Unit, TUM; (E) common ragweed (*Ambrosia artemisiifolia*), native to North America, © Bodner, Southern Weed Science Society, Bugwood.org; (F) signal crayfish (*Pacifastacus leniusculus*), native to North America, © Aquatic Systems Biology Unit, TUM.

Terrestrial Plants

Many invasive plant species in Europe are primarily recognized as agricultural or forestry weeds. Additionally, 17 out of the 18 plant species recorded among the most damaging invasive species in Europe [6] are known to reduce the habitat of native species [34].

Eight of them are reported to disrupt community assemblages, for example by impacting plant pollinator networks [34]. Non-native plant species can also hybridize with closely related native species so that distinctive genotypes of native plants are lost [102]. Species such as Japanese knotweed (*Fallopia japonica*) and Himalayan balsam (*Impatiens glandulifera*) grow and are nuisance species along railway lines (the former) and waterways (both). Other plant species can cause severe health problems. For example, giant hogweed (*Herracleum mantegazzianum*, Figure 1C) produces sap that causes skin lesions to humans upon contact [103]. The pollen of invasive ragweed (*Ambrosia artemisiifolia*, Figure 1E) is highly allergenic to humans, and estimates of associated medical costs in Germany range between 17 and 47 million EUR annually [104]. According to Vilà et al. [7], the most costly plant invaders affecting nature conservation, agriculture, forestry, and fisheries in Europe are pigface species (*Carpobrotus*spp.). These produce annual costs for control and eradication in Spain of approximately 0.58 million EUR [105]. Overall, invasive terrestrial plants cause costs of at least 3.7 billion EUR per year in Europe [99].

Aquatic Organisms

Invasive species are considered one of five major threats to aquatic biodiversity worldwide [4], with particularly large impacts on freshwater habitats [106,107]. The isolated nature of most freshwater habitats means that natural spread of aquatic organisms into new habitats occurs at low frequencies. In turn, this means that aquatic communities tend to be more different to each other, and thus that the increased rates of species movement caused by human pathways have large potential for impacts on biodiversity.

Traditionally, the study of invasions in aquatic ecosystems has had a strong focus on economically important and visible species, whereas invasive populations of small taxa (e.g. plankton) or groups that are difficult to identify at the species level (e.g. chironomids) have rarely been considered. As a larger and more visible species, North American signal crayfish (*Pacifastacus leniusculus*, Figure 1F) have been relatively well monitored and recorded. They were introduced to Europe primarily for aquaculture, have spread rapidly, and are now considered one of the major threats to the indigenous crayfish fauna [108]. In addition to their competitive behavior [109], North American crayfish are hosts of the crayfish plague (*Aphanomyces astaci*), an oomycete fungus that causes a lethal disease to European crayfish [110]. Also, the introduction of non-native salmonids and gobiids (e.g.*Neogobius melanostomus*, Figure 1D) has resulted in the decline and even extinction of indigenous species, and caused ecosystem shifts in lakes and streams [6,111]. In economic terms, the zebra mussel *Dreissena polymorpha*, which can completely block cooling systems in hydropower plants, probably has the greatest impact of all freshwater invaders.

In marine habitats, negative effects of non-native species include declines in native species richness and abundance. These impacts have been associated with the invasion of *Caulerpa taxifolia* into the Mediterranean [112], and with the high mortality rates of European oysters (*Ostrea edulis*) due to competition with introduced Pacific oysters (*Crassostrea gigas*) and damage from introduced parasites. Despite these examples, there is little comprehensive evidence for most impacts of invasive marine species, and there are some examples of economic benefits. For example, the release of the red king crab (*Paralithodes camtschaticus*) into the Barents Sea and its southward spread along the Norwegian coast has provided an additional fishery and income for fishermen in the order of 9 million EUR per year [36]. Nonetheless, negative impacts of invasive aquatic species in Europe are high and have been estimated to cost at least 2.2 billion EUR per year [99].

MANAGEMENT AND POLICY FOR INVASIVE SPECIES

Policy Challenges

Compared to other environmental problems, invasive species present at least six particular challenges. First, their impacts tend to increase over time as populations become larger and spread [113]. In contrast, other environmentally damaging activities, such as the release of chemical pollutants, generally decrease in severity over time after the activity is ended. This means that populations of invasive species can best be managed through rapid eradication of new populations [114]. This is most feasible soon after a species becomes established, but the type of biological information needed to support eradication and the resources and political will to eradicate a harmful invader are generally not available until the species has spread and become invasive. By this point there are fewer options to control the population, and the level of resources required, and the possibility for undesirable side effects on other ecosystem components, are usually prohibitive.

Second, many other issues for environmental policy, such as forestry and mineral exploration, can be effectively managed with little concern for the policies followed by one's neighbours. In contrast, invasive species readily cross borders as long as there is suitable habitat on the other side [1]. This means that efforts to prevent invasions need to be mindful of the efforts being made in neighbouring jurisdictions [1]. As European economies become more integrated there are fewer barriers, such as border checks, to the spread of invasive species.

Third, invasive species often travel as contaminants of valuable trade. For example, most large ships carry ballast water to increase their weight when they are not fully loaded with cargo. This ballast is taken on at one port, along with any organisms in the water, and discharged at subsequent ports as cargo is loaded. Unintentional transfer of organisms in ballast tanks is now the largest single cause of marine invasions [115], and a strong factor in freshwater invasions [28,116]. Reducing the spread of invasive species via this pathway would either require constraints on where ships travel, or the installation onto all ships of expensive ballast treatment technology. Either of these options would increase the cost of shipped goods. Thus, control of invasive species often involves trade-offs with other activities, complicating decisions about how impacts can best be managed [114].

Fourth, it is usually impossible to determine the exact conditions that will lead to an invasion. Risky activities, such as collecting ballast water in one port and dumping it in another, may occur many times before an invader becomes established. The invasion of the Laurentian Great Lakes by zebra mussels (*Dreissena polymorpha*) in the late 1980s, for example, had been predicted for over a century before it actually occurred [117]. These predictions were based on shipping connections between the Great Lakes and regions where zebra mussel was established, and on the fact that the species is well suited to Great

Lakes conditions. Why it took so long for the invasion to occur and why it occurred when it did remains unknown. This general uncertainty about invasions complicates management because economic activities with definite and well-quantified immediate benefits must be balanced against risks from the stochastic invasion process [118].

Fifth, controlling the spread of invasive species requires international cooperation, and this is often difficult to achieve [1]. For example, in response to the risk from ballast water invasions, the International Maritime Organization (IMO) produced the *International Convention for the Control and Management of Ships Ballast Water and Sediments* in February 2004. This convention requires the establishment of ballast water management systems on ships, with the goal of preventing the movement of live organisms. The dates at which each ship would be required to have ballast water treatment facilities differ based on ship size and age, but fall between 2009 and 2016. Despite this convention being finalized for several years and despite a widespread consensus that its implementation could reduce rates of invasion, it has not yet come into effect because too few countries have ratified it. Many other pathways of species invasion could be best managed at an international level, but policies have rarely been negotiated and implemented at this level [1].

Finally, there is often a considerable time lag, occasionally as much as several centuries, between the introduction of non-native species and their spread. These time lags appear to be especially long for terrestrial plants [12]. In a recent paper, Essl et al. [13] have shown that this may lead to an 'invasion debt' in which human activities many decades ago may have already set in process invasions that have not yet been recorded. In their paper, Essl et al. [13] showed that for several groups of plants and animals the socio-economic conditions in Europe during 1900 were more closely related to current levels of invasions than conditions in the year 2000, even though many species were introduced in the second half of the twentieth century. This means that current activities in Europe may be setting in motion many future invasions.

Bioeconomic Approaches to Policy and Management

Ecological studies of species invasions have increased in number and sophistication over the last several decades [19]. This has created a much clearer picture of the number of species that are established beyond their native range, and the types and severity of impacts that they cause. Recognition of these impacts has led to increased calls for management and policy (e.g. [119]). Although many nations now have policies to reduce the impacts of invasive species, invaders continue to arrive at an increasing rate, indicating that the results from ecological studies have not prompted sufficient changes in policy and management.

Concurrent with advances by ecologists, economists have taken an interest in the problems of invasive species. Much of their interest centres on the trade-offs that must be made to manage invasive species [114]. For example, economists might make the link between a beneficial trade route and the risks from invasions that it causes. They can then ask if and how the trade route can be restricted or modified to reduce risks from invasive species while maximizing overall societal welfare (e.g. [120]).

Although ecologists and economists have often addressed similar concerns (i.e. how to reduce the impacts of invasive species), they have rarely worked together. This stems in part from a lack of understanding between the disciplines. In particular, ecologists may be unwilling to consider ecosystem components/qualities (e.g. biodiversity) in monetary terms. On the economist side, there is often a lack of willingness to work with ecological systems because they are so complex, in many cases are poorly understood even by ecologists, and because the relationships among ecosystem components may not fit into traditional economic models.

As described above, the introduction of non-native species is a consequence of the globalization of trade and travel. The benefits to society from globalization are large, as are the benefits from non-native species used in agriculture, horticulture, forestry, fisheries, landscaping, and the pet trade, etc. Thus, the benefits from policy that prevents and manages invasions must be balanced against the costs of the policy, including the costs of restricting trade. In practical terms, this requires a good understanding of both the ecological and economic dimensions of invasions, and requires that economists and ecologists work together to design appropriate policies.

General Principles for Policy and Management

It is generally accepted, and has often been shown, that the best way to reduce total impacts from invasive species is to prevent their introduction [113,119,121]. For example, studies have shown that preventing the spread of an invader among lakes [122,123], and the spread of species among countries through trade [120], is economically preferable to managing species after they arrive. If prevention of all introductions is not feasible, efforts to reduce propagule pressure should be the goal as this may effectively prevent a proportion of invasions. Prevention is considered more effective than managing established invaders for at least three reasons. First, it is the only sure way to avoid impacts of invasive species, which as noted above are rarely eradicated. Second, there are more management options available prior to invasion. These include a variety of laws and regulations, quarantine efforts, and the threat of punitive action against individuals or companies that do not comply. Finally, prevention efforts give the opportunity to shift the costs of reducing impacts to the trades that derive greatest benefits from the movement of species [1]. For example, although the pet trade has introduced many harmful invaders around the world, the exporters, importers, and retailers who derive greatest benefits from the trade have not been required to pay the costs. Instead, those costs are generally borne by society as government agencies work to control invaders. Prior to invasion, however, it is possible to impose regulations that require the trades to ensure that they are not transporting potential invaders. Thus, the externality of trade can be internalized.

Post-establishment, the broad range of invasive species and the ecosystems they become established in, and the heterogeneous ways that humans value those ecosystems, means that few generalizations can be given for management and policy. That is, appropriate management and policy for harmful invasive species is highly context dependent. Options include poisoning (e.g. herbicides, pesticides, piscicides), manual removal (e.g. pulling plants), capturing and killing (e.g. trapping, shooting), and release of additional species that may provide biological control of the invader. As the applications of these methods are context dependent, and because different countries within Europe have different standards for the application of these methods, they are not explored further here. The general principle of prevention still applies, however, and efforts to slow or eliminate population spread will often be the most cost-effective ways to reduce total impacts.

INVASIVE SPECIES POLICY IN EUROPE

There is a large diversity of national approaches to invasive species within the EU (see review in [124]). These range from relatively advanced policies and procedures, such as in the Netherlands where there is a national policy framework for addressing invasive species, to nations that do relatively little to prevent the arrival and spread of invaders. Open internal borders, however, create a weak-link problem in the EU where even nations with the most stringent policies remain at risk because some of their neighbors and trading partners are doing relatively little [1].

At the international level there are several instruments that address the threats from invasive species. These include the Convention on Biological Diversity (CBD, implemented into European legislation by Council Decision 93/626/EEC), which requires ratifying nations to work to prevent the introduction, spread, and export of all types of invasive species (http://www.cbd.int/). There is no history, however, of nations being sanctioned for not following the invasive species guidelines, and there is little evidence that the CBD has led to improved invasive species management at the global scale.

Non-native species that pose direct threats to animals or plants (cultivated or wild) are actively managed at the international level by the World Organization for Animal Health (OIE) and International Plant Protection Convention (IPPC), respectively. While these instruments have proven effective for restricting the spread of some invaders, they have most often been used to address diseases of livestock and pests of crops, with less attention paid to invasive species that cause primarily environmental harm [124]. Within the framework of the IPPC are the European and Mediterranean Plant Protection Convention which develops regional measures to prevent the spread of invasive plants and plant pests for its 50 member nations (i.e. more than just the EU nations). Its focus has also been largely on pests of agriculture.

Although the EU now has policies that address several region-wide environmental issues (e.g. climate change, pollution by chemicals), it has not yet created legislation to address the full range of invasive species and their pathways. Instead, there are several regimes and directives that address some invasive species or some pathways of introduction. We give two examples here. First, Council Regulation 708/2007 sets procedures and standards for aquaculture within the EU with the goal of reducing the risks that non-native fish, or the diseases they carry, will become established and invasive. Second, the Plant Health Regime addresses the introduction and spread of invasive plants and plant pests, but has been used primarily to address pests of agriculture. Plants that pose risks of environmental harm, although they may arrive through the same pathways, are generally not covered.

Although the two examples just given address important aspects of the invasive species problem, the general European policy framework remains spread across many similar and disjunct pieces of legislation. This creates confusion about which species and pathways should be addressed by which instrument, and ultimately means that many invasive species and pathways are not addressed [124]. In response, several programs have been established to identify gaps in the current framework and to suggest solutions. One example is the IMPASSE program that has reviewed the risks of introduction of invasive fish species and fish diseases through aquaculture (http://ec.europa.eu/research/fp6/ssp/impasse_en.htm). However, even if all recommendations from these programs were enacted, the basic European legislative framework would remain scattered, creating large obstacles to a coordinated response to the invasive species problem.

This lack of coordination has been identified as a serious shortcoming by the Commission of the European Communities [14]. In response, the commission set forth three policy options. The first is a business as usual model, in which the EU continues to rely on the current suite of national and EU policies, and the range of international policy instruments, to address invasive species. The second is to maximize use of the existing EU policy framework, possibly with some modifications of legal instruments so that they better address invasive species. This approach appeals because it would not require new legislation. However, it would still be limited in its application because no current policies address all invasive species or pathways. The final option is to create a new legal instrument at the EU level that would apply to all member states. This third option could be complemented by the creation of a European Centre for Invasive Species Management that would provide a coordinating role [125]

Shine et al [124] recently prepared a comprehensive review of existing national, international and European policy, and built upon this to evaluate the three policy options. They strongly recommended the final option - a new legislative instrument - and noted that it is the only option that would provide a basis for coordinated action against all invasive species. Additionally, an economic analysis showed that although this option would be the most expensive, its costs would be much less than the avoided costs of the invasive species that it would prevent from arriving, or that would be managed more efficiently [124]. This recommendation is currently under consideration by the EU, and a final decision about policy approach is expected by 2012.

CONCLUSIONS

Europe's position as a centre for international trade over many centuries has resulted in it having a large number and diversity of established non-native species. Many of these species are invasive and they affect all European habitats. Impacts include loss of native biodiversity, economic losses, and harm to human health. Over the last several decades the field of invasion biology has grown markedly [19] and has created a growing understanding of the biology of invasive species. One of the emergent insights of invasion biology has been that invasions are an inherently multi-disciplinary issue. More specifically, species are introduced through human economic activities. Without a clear understanding of these activities it is impossible to explain and understand biological invasions. This is clearly illustrated by the importance of propagule pressure as a driver of invasions. Propagule pressure is driven by human activities and is increasingly recognized as a strong factor in many, if not most, invasions.

Where timelines of species non-native species discovery are available, it is clear that the problem of invasions has not been fully addressed by current national or EU policies. Indeed, many European habitats appear to be acquiring new non-native species faster now than at any time in recorded history. Creating and enforcing policies that slow the rate of invasion is difficult for many reasons, including the value derived from many of the pathways that move the most species (e.g. shipping). These policies may be especially difficult in Europe because it is an environmentally heterogeneous region where barriers to the movement of goods have been actively removed to encourage free trade and travel. Despite this difficulty, current deliberations in the EU about developing a comprehensive policy instrument to address invasive species are very promising.

Apart from the policy and management considerations, the spread of non-native species offers many insights into biological and ecological processes as well as into the links between biodiversity, ecosystem functioning, and socio-economic impacts. Many researchers have been attracted to study species invasions because of the complexities, and the possibility to integrate across disciplines in ways that both generate new insight and that can lead to practical policy solutions. By reviewing a broad range of invasive species, and by reviewing both the biological and economic literature, we hope that this manuscript can make a useful contribution toward a more integrative understanding of species invasions in Europe.

AUTHORS' CONTRIBUTIONS

RK, JG, JJ, and IK conceived of and designed this review paper. RK, JG, JJ, and IK each contributed research and draft text sections. RK organized all text into the final version. JG, JJ, and IK contributed equally. All authors read and approved the final manuscript.

ACKNOWLEDGEMENTS

We thank both the Alexander von Humboldt Foundation and the U. S. National Academy of Sciences for bringing us together at the 16th German-American Frontiers of Science Symposium in Potsdam, Germany, June 2010. We especially thank Henner Hollert for inviting us to write this special review paper, and appreciate helpful comments by Ingo Kowarik and two anonymous reviewers. JMJ acknowledges financial support from the Deutsche Forschungsgemeinschaft (DFG; JE 288/4-1).

REFERENCES

1. Perrings C, Burgiel S, Lonsdale WM, Mooney H, Williamson M: International cooperation in the solution to trade-related invasive species risks. *Ann NY Acad Sci* 2010, 1195:198-212.
2. Kolar CS, Lodge DM: Progress in invasion biology: predicting invaders. *Trends Ecol Evol* 2001, 16:199-204.
3. Williamson M: *Biological invasions*. London: Chapman & Hall; 1996.
4. Sala OE, Chapin FS, Armesto JJ, Berlow E, Bloomfield J, Dirzo R, Huber-Sanwald E, Huenneke Lf, Jackson RB, Kinzig A, Leemans R, Lodge DM, Mooney HA, Oesterheld M, Poff NL, Sykes MT, Walker BH, Walker M, Wall DH: Global Biodiversity Scenarios for the Year 2100. *Science* 2000, 287:1770-1774.
5. Jeschke JM, Strayer DL: Invasion success of vertebrates in Europe and North America. *P Natl Acad Sci USA* 2005, 102:7198-7202.
6. DAISIE: *Handbook of alien species in Europe*. Dordrecht: Springer; 2009.
7. Vilà M, Basnou C, Pyšek P, Josefsson M, Genovesi P, Gollasch S, Nentwig W, Olenin S, Roques A, Roy D, Hulme PE, DAISIE partners: How well do we understand the impacts of alien species on ecosystem services? A pan-European cross-taxa assessment. *Front Ecol Environ* 2010, 8:135-144.
8. Keller RP, zu Ermgassen PSE, Aldridge D: Vectors and timing of nonindignenous freshwater species establishment in Great Britain. *Conserv Biol* 2009, 23:1526-1534.
9. Hulme PE, Bacher S, Kenis M, Klotz S, Kühn I, Minchin D, Nentwig W, Olenin S, Panov V, Pergl J, Pyšek P, Roques A, Sol D, Solarz W, Vilà M: Grasping at the routes of biological invasions: a framework for integrating pathways into policy. *J Appl Ecol* 2008, 45:403-414.

10. Lambdon PW, Pyšek P, Basnou C, Hejda M, Arianoutsou M, Essl F, Jarosik V, Pergl J, Winter M, Anastasiu P, Andriopoulos P, Bazos I, Brundu G, Celesti-Grapow L, Chassot P, Delipetrou P, Josefsson M, Kark S, Klotz S, Kokkoris Y, Kühn I, Marchante H, Perglova I, Pino J, Vilà M, Zikos A, Roy D, Hulme PE: Alien flora of Europe: species diversity, temporal trends, geographical patterns and research needs. *Preslia* 2008, 80:101-149.
11. Kowarik I: *Biologische Invasionen: Neophyten und Neozoen in Mitteleuropa*. 2nd edition. Stuttgart: Ulmer; 2010.
12. Kowarik I: Time lags in biological invasions with regard to the success and failure of alien species. In *Plant Invasions: General Aspects and Special Problems*. Edited by Pyšek P, Prach K, Rejmánek M, Wade M. Amsterdam: SPB Academic Publishing; 1995:15-38.
13. Essl F, Dullinger S, Rabitsch W, Hulme PE, Huelber K, Jarošík V, Kleinbauer I, Krausmann F, Kühn I, Nentwig W, Vilà M, Genovesi P, Gherardi F, Desprez-Loustau ML, Roques A, Pyšek P:Socio-economic legacy yields an invasion debt. *P Natl Acad Sci USA* 2011, 108:203-207.
14. Commission of the European Communities: Towards a strategy on invasive species.[http:/ / ec.europa.eu/ environment/ nature/ invasivealien/ docs/ 1_EN_ACT_part1_v6.pdf]*webcite* 2008. Communication from the Commission to the Council, the European Parliament, the European Economic and Social Committee and the Committee of Regions. COM(2008) 789 final
15. Council of the European Union: A mid-term assessment of implementing the EU Biodiversity Action Plan and Towards an EU Strategy on Invasive Alien Species – Council conclusions. [http://register.consilium.europa.eu/pdf/en/09/st11/st11412.en09.pdf] *webcite* 2009. Information Note 11412/09
16. Lodge DM: Biological invasions: lessons for ecology. *Trends Ecol Evol* 1993, 8:133-137.
17. Sax DF, Stachowicz JJ, Brown JH, Bruno JF, Dawson MN, Gaines SD, Grosberg RK, Hastings A, Holt RD, Mayfield MM, O'Connor MI, Rice WR: Ecological and evolutionary insights from species invasions. *Trends Ecol Evol* 2007, 22:465-471.
18. Richardson DM, (ed): *Fifty Years of Invasion Ecology - The Legacy of Charles Elton*. Chichester (UK): Wiley-Blackwell; 2011.

19. Richardson DM, Pyšek P, Rejmánek M, Barbour MG, Panetta FD, West CJ: Naturalization and invasion of alien plants: concepts and definitions. *Divers Distrib* 2000, 6:93-107.
20. Colautti RI, MacIsaac HJ: A neutral terminology to define 'invasive' species. *Divers Distrib* 2004, 10:135-141.
21. Jeschke JM: Across islands and continents, mammals are more successful invaders than birds. *Divers Distrib* 2008, 14:913-916.
22. Thellung A: La flore adventice de Montpellier. *Mémoires de la Société nationale des Sciences naturelles et mathématiques de Cherbourg*1912, 38:57-728.
23. Thellung A: Pflanzenwanderungen unter dem Einfluss des menschen. *Beibl Bot jahrb Syst* 1915, 53:37-66.
24. Wilson JRU, Dormontt EE, Prentis PJ, Lowe AJ, Richardson DM: Something in the way you move: dispersal pathways affect invasion success. *Trends Ecol Evol* 2009, 24:136-144.
25. Kowarik I, von der Lippe M: Pathways in plant invasions. In *Biological Invasions*. Edited by Nentwig W. Berlin: Springer; 2007:29-47.
26. Holdich DM, Rogers WD, Reynolds JD: Native and alien crayfish in the British Isles. In*Crayfish in Europe as alien species: how to make the best of a bad situation?*. Edited by Gherardi F, Holdich DM. Rotterdam: AA Balkema; 1999:221-235.
27. Benedict MQ, Levine RS, Hawley WA, Lounibos LP: Spread of the tiger: global risk of invasion by the mosquito *Aedes albopictus*. *Vector-Borne Zoonot* 2007, 7:76-85.
28. Keller RP, Drake JM, Drew M, Lodge DM: Linking environmental conditions and ship movements to estimate invasive species transport across the global shipping network. *Divers Distrib* 2011, 17:93-102.
29. Olenin S, Minchin D, Daunys D, Ziako A: Pathways of aquatic invasions in Europe. In*Atlas of Biodiversity Risk*. Edited by Settele J, Penev LD, Georgiev TA, Grabaum R, Grobelnik V, Hammen V, Klotz S, Kotarac M, Kühn I. Sofia (Bulgaria): Pensoft; 2010:138-139.
30. Augustin S, Guichard S, Heitland W, Freise J, Svatoš A, Gilbert M: Monitoring and dispersal of the invading Gracillariidae *Cameraria ohridella*. *J Appl Entomol* 2009, 133:58-66.

31. Genovesi P, Bacher S, Kobelt M, Pascal M, Scalera R: Alien mammals of Europe. In*Handbook of Alien Species in Europe*. Edited by DAISIE. Dordrecht: Springer; 2009:119-128.

32. Kark S, Solarz W, Chiron F, Clergeau P, Shirley S: Alien birds, amphibians and reptiles of Europe. In *Handbook of Alien Species in Europe*. Edited by DAISIE. Dordrecht: Springer; 2009:105-118.

33. Roques A, Rabitsch W, Rasplus JY, Lopez-Vaamonde C, Nentwig W, Kenis M: Alien terrestrial invertebrates of Europe. In *Handbook of Alien Species in Europe*. Edited by DAISIE. Dordrecht: Springer; 2009:63-79.

34. Pyšek P, Lambdon PW, Arianoutsou M, Kühn I, Pino J, Winter M: Alien vascular plants of Europe. In *Handbook of Alien Species in Europe*. Edited by DAISIE. Dordrecht: Springer; 2009:43-61.

35. Bourke PMA: Emergence of potato blight, 1843-46. *Nature* 1964, 203:805-808.

36. Galil BS, Gollasch S, Minchin D, Olenin S: Alien marine biota of Europe. In *Handbook of Alien Species in Europe*. Edited by DAISIE. Dordrecht: Springer; 2009:93-104.

37. Gherardi F, Gollasch S, Minchin D, Olenin S, Panov VE: Alien invertebrates and fish in European inland waters. In *Handbook of Alien Species in Europe*. Edited by DAISIE. Dordrecht: Springer; 2009:81-92.

38. Leuven RSEW, van der Velde G, Baijens I, Snijders J, van der Zwart C, Lenders HJR, bij de Vaate A: The river Rhine: a global highway for dispersal of aquatic invasive species. *Biol Invasions* 2009, 11:1989-2008.

39. Sol D, Vilà M, Kühn I: The comparative analysis of historical alien introductions. *Biol Invasions* 2008, 10:1119-1129.

40. Darwin C: *On the Origin of Species by Means of Natural Selection, or the Preservation of Favoured Races in the Struggle for Life*. London: Murray; 1859.

41. Cadotte MW: Darwin to Elton: early ecology and the problem of invasive species. In*Conceptual Ecology and Invasion Biology: Reciprocal Approaches to Nature*. Edited by Cadotte MW, McMahon SM, Fukami T. Dordrecht: Springer; 2006:15-33.

42. Elton CS: *The ecology of invasions by animals and plants.* London: Methuen; 1958.
43. Levine JM, D'Antonio CM: Elton revisited: A review of evidence linking diversity and invasibility. *Oikos* 1999, 87:15-26.
44. Deutschewitz K, Lausch A, Kühn I, Klotz S: Native and alien plant species richness in relation to spatial heterogeneity on a regional scale in Germany. *Global Ecol Biogeogr* 2003, 12:299-311.
45. Stohlgren TJ, Barnett DT, Kartesz JT: The rich get richer: patterns of plant invasions in the United States. *Front Ecol Environ* 2003, 1:11-14.
46. Fridley JD, Stachowicz JJ, Naeem S, Sax DF, Seabloom EW, Smith MD, Stohlgren TJ, Tilman D, Von Holle B: The invasion paradox: Reconciling pattern and process in species invasions. *Ecology* 2007, 88:3-17.
47. Kühn I, Klotz S: From Ecosystem invasibility to local, regional and global patterns of invasive species. In *Biological Invasions.* Edited by Nentwig W. Heidelberg: Springer; 2007:181-196.
48. Duncan RP, Blackburn TM, Sol D: The ecology of bird introductions. *Annu Rev Ecol Evol S* 2003, 34:71-98.
49. Leprieur F, Beauchard O, Blanchet S, Oberdorff T, Brosse S: Fish invasions in the world's river systems: when natural processes are blurred by human activities. *PLoS Biol* 2008, 6:e28. 2008
50. Blackburn TM, Cassey P, Lockwood JL: The island biogeography of exotic bird species. *Global Ecol Biogeogr* 2008, 17:246-251.
51. Chiron F, Shirley S, Kark S: Human-related processes drive the richness of exotic birds in Europe. *P Roy Soc B-Biol Sci* 2009, 276:47-53.
52. Jeschke JM, Genovesi P: Do biodiversity and human impact influence the introduction or establishment of alien mammals? *Oikos* 2011, 120:57-64.
53. Kühn I, May R, Brandl R, Klotz S: Plant distribution patterns in Germany - Will aliens match natives? *Feddes Repertorium* 2003, 114:559-573.
54. Pino J, Font X, Carbo J, Jove M, Pallares L: Large-scale correlates of alien plant invasion in Catalonia (NE of Spain). *Biol Conserv* 2005, 122:339-350.

55. Pyšek P, Kucera T, Jarosik V: Plant species richness of nature reserves: the interplay of area, climate and habitat in a central European landscape. *Global Ecol Biogeogr* 2002, 11:279-289.
56. Gasso N, Sol D, Pino J, Dana ED, Lloret F, Sanz-Elorza M, Sobrino E, Vilà M: Exploring species attributes and site characteristics to assess plant invasions in Spain. *Divers Distrib* 2009, 15:50-58.
57. Chytrý M, Pyšek P, Wild J, Pino J, Maskell LC, Vilà M: European map of alien plant invasions based on the quantitative assessment across habitats. *Divers Distrib* 2009, 15:98-107. 2009
58. Davis MA, Grime JP, Thompson K: Fluctuating resources in plant communities: a general theory of invasibility. *J Ecol* 2000, 88:528-534.
59. Britton-Simmons KH, Abbott KC: Short- and long-term effects of disturbance and propagule pressure on biological invasion. *J Ecol* 2008, 96:68-77.
60. di Castri F: History of biological invasions with special emphasis on the Old World. In*Biological Invasions: A Global Perspective.* Edited by Drake JA, Mooney HA, di Castri F, Groves RH, Kruger FJ, Rejmánek M, Williamson M. Chichester UK: John Wiley and Sons; 1989:1-30.
61. Becker T, Dietz H, Billeter R, Buschmann H, Edwards PJ: Altitudinal distribution of alien plant species in the Swiss Alps. *Perspect Plant Ecol* 2005, 7:173-183.
62. Pyšek P, Jarošík V, Hulme PE, Kühn I, Wild J, Arianoutsou M, Bacher S, Chiron F, Didžiulis V, Essl F, Genovesi P, Gherardi F, Hejda M, Kark S, Lambdon PW, Desprez-Loustau ML, Nentwig W, Pergl J, Poboljšaj K, Rabitsch W, Roques A, Roy DB, Shirley S, Solarz W, Vilà M, Winter M:Disentangling the role of environmental and human pressures on biological invasions across Europe. *P Natl Acad Sci USA* 2010, 107:12157-12162.
63. Galil BS: A sea under siege - alien species in the Mediterranean. *Biol Invasions* 2000, 2:177-186.
64. Leppäkoski E, Olenin S, Gollasch S: The Baltic Sea: a field laboratory for invasion biology. In *Invasive Aquatic Species of Europe: Distribution, Impacts and Management.* Edited by Leppäkoski E, Gollasch S, Olenin S. Dordrecht: Kluwer; 2002:253-259.

65. Panov VE, Bychenkov DE, Berezina NA, Maximov AA: Alien species introductions in the eastern Gulf of Finland: current state and possible management options. *Proc Est Acad Sci Biol Ecol* 2003, 52:254-267.

66. Panov V, Dgebuadze Y, Shiganova T, Filippov A, Minchin D: A risk assessment of biological invasions: inland waterways of Europe - the northern invasion corridor case study.In *Biological Invaders in Inland Waters: Profiles, Distribution, and Threats*. Edited by Gherardi F. Dordrecht: Springer; 2007:639-656.

67. Wolff WJ, Reise K: Oyster imports as a vector for the introduction of alien species into northern and western European coastal waters. In *Invasive Aquatic Species of Europe: Distribution, Impacts and Management*. Edited by Leppäkoski E, Gollasch S, Olenin S. Dordrecht: Kluwer; 2002:193-205.

68. Baker HG: The evolution of weeds. *Annu Rev Ecol Syst* 1974, 5:1-24.

69. Jeschke JM, Strayer DL: Determinants of vertebrate invasion success in Europe and North America. *Global Change Biol* 2006, 12:1608-1619.

70. Pyšek P, Richardson DM: Traits Associated with Invasiveness in Alien Plants: Where Do we Stand? In *Biological Invasions*. Edited by Nentwig W. Dordrecht: Springer; 2007:97-125.

71. Genovesi P, Shine C: *European strategy on invasive alien species*. Strasbourg: Council of Europe Publishing; 2004.

72. Weber E, Köhler B, Gelpke G, Perrenoud A, Gigon A: Schlüssel zur Einteilung von Neophyten in der Schweiz in die Schwarze Liste oder die Watch-Liste. *Bot Helv* 2005, 115:169-194.

73. Gederaas L, Salvesen I, Viken Å: *Norwegian Black List 2007: Ecological risk analysis of alien species*. Trondheim: Artsdatabanken; 2007.

74. Parrott D, Roy S, Baker R, Cannon R, Eyre D, Hill M, Wagner M, Preston C, Roy H, Beckmann B, Copp GH, Edmons N, Ellis J, Laing I, Britton R, Gozlan RE, Mumford J: *Horizon scanning for new invasive non-native animal spe-cies in England*. Sheffield: Natural England; 2009.

75. Nehring S, Essl F, Klingenstein F, Nowack C, Rabitsch W, Stöhr O, Wiesner C, Wolter C:*Schwarze Liste invasiver Arten:*

Kriteriensystem und Schwarze Listen invasiver Fische für Deutschland und für Österreich. Bonn (Germany): Bundesamt für Naturschutz; 2010.

76. Keller RP, Drake JM: Trait based risk assessment for invasive species. In *Bioeconomics of Invasive Species: Integrating Ecology, Economics, Policy and Management*. Edited by Keller RP, Lodge DM, Lewis MA, Shogren JF. New York: Oxford University Press; 2009:44-62.
77. Hayes KR, Barry SC: Are there any consistent predictors of invasion success? *Biol Invasions* 2008, 10:483-506.
78. Lockwood JL, Cassey P, Blackburn T: The role of propagule pressure in explaining species invasions. *Trends Ecol Evol* 2005, 20:223-228.
79. Blackburn TM, Duncan RP: Establishment patterns of exotic birds are constrained by non-random patterns in introduction. *J Biogeogr* 2001, 28:927-939.
80. Sol D, Duncan RP, Blackburn TM, Cassey P, Lefebvre L: Big brains, enhanced cognition, and response of birds to novel environments. *P Natl Acad Sci USA* 2005, 102:5460-5465.
81. Cassey P, Blackburn T, Russell GJ, Jones KE, Lockwood JL: Influences on the transport and establishment of exotic bird species: an analysis of the parrots (Psittaciformes) of the world. *Global Change Biol* 2004, 10:417-426.
82. Blackburn TM, Cassey P, Lockwood JL: The role of species traits in the establishment success of exotic birds. *Global Change Biol* 2009, 15:2852-2860.
83. Hanspach J, Kühn I, Pyšek P, Boos E, Klotz S: Correlates of naturalization and occupancy of introduced ornamentals in Germany. *Perspect Plant Ecol* 2008, 10:241-250.
84. Williamson M, Dehnen-Schmutz K, Kühn I, Hill M, Klotz S, Milbau A, Stout J, Pyšek P: The distribution of range sizes of native and alien plants in four European countries and the effects of residence time. *Divers Distrib* 2009, 15:158-166.
85. Pyšek P, Jarošik V, Pergl J, Randall R, Chytrý M, Kühn I, Tichy L, Danihelka J, Jun JC, Sádlo J:The global invasion success of Central European plants is related to distribution characteristics in their native range and species traits. *Divers Distrib* 2009, 15:891-903. 2009a

86. van Kleunen M, Weber E, Fischer M: A meta-analysis of trait differences between invasive and non-invasive plant species. *Ecol Lett* 2010, 13:235-245.

87. Dawson W, Fischer M, van Kleunen M: The maximum relative growth rate of common UK plant species is positively associated with their global invasiveness. *Global Ecol Biogeogr* 2011, 20:299-306.

88. Küster EC, Kühn I, Bruelheide H, Klotz S: Trait interactions help explain plant invasion success in the German flora. *J Ecol* 2008, 96:860-868.

89. Thompson K, Davis MA: Why research on traits of invasive plants tells us very little. *Trends Ecol Evol* 2011, 26:155-156.

90. Paavola M, Olenin S, Leppäkoski E: Are invasive species most successful in habitats of low native species richness across European brackish water seas? *Estuar Coast Shelf S* 2005, 64:738-750.

91. Strayer D, Downing JA, King TL, Layer JB, Newton TJ, and Nichols SJ: Changing perspectives on pearly mussels, North America's most imperiled animals. *BioScience* 2004, 54:429-439.

92. Geist J: Strategies for the conservation of endangered freshwater pearl mussels (*Margaritifera margaritifera* L.): a synthesis of Conservation Genetics and Ecology. *Hydrobiologia* 2010, 644:69-88.

93. Keller RP, Drake JM, Lodge DM: *Fecundity as a basis for risk assessment of nonindigenous freshwater molluscs. Conserv Biol.* 2007, 21:191-200.

94. Karatayev AY, Burlakova LE, Padilla DK, Matitsky SE, Olenin S: Invaders are not a random selection of species. *Biol Invasions* 2009, 11:2009-2019.

95. Strayer DL, Eviner VT, Jeschke JM, Pace ML: Understanding the long-term effects of species invasions. *Trends Ecol Evol* 2006, 21:645-651.

96. Bisby FA, Roskov YR, Orrell TM, Nicolson D, Paglinawan LE, Bailly N, Kirk PM, Bourgoin T, Baillargeon G: Species 2000 & ITIS Catalogue of Life: 2010 Annual Checklist.[http://www.catalogueoflife.org/annual-checklist/2010] *webcite* Accessed 30 July 2010

97. Pyšek P, Richardson DM, Rejmánek M, Webster BL, Williamson M, Kirschner J: Alien plants in checklists and floras: towards better communication between taxonomists and ecologists. *Taxon* 2004, 53:131-143.

98. Gherardi F, Bertolino S, Bodon M, Casellato S, Cianfanelli S, Ferraguti M, Lori E, Mura G, Nocita A, Riccardi N, Rossetti G, Rota E, Scalera R, Zerunian S, Tricarico E: Animal xenodiversity in Italian inland waters: distribution, modes of arrival, and pathways. *Biol Invasions* 2008, 10:435-454.

99. Kettunen M, Genovesi P, Gollasch S, Pagad S, Starfinger U, ten Brink P, Shine C: *Technical support to EU strategy on invasive alien species (IAS) - Assessment of the impacts of IAS in Europe and the EU.* Brussels: Institute for European Environmental Policy; 2009.

100. Kumschick S, Nentwig W: Some alien birds have as severe an impact as the most effectual alien mammals in Europe. *Biol Conserv* 2010, 143:2757-2762.

101. Nentwig W, Kühnel E, Bacher S: A generic impact-scoring system applied to alien mammals in Europe. *Conserv Biol* 2010, 24:302-311.

102. Vilà M, Weber E, D'Antonio CM: Conservation implications of invasion by plant hybridization. *Biol Invasions* 2000, 2:207-217.

103. Pyšek P, Cock MJW, Nentwig W, Ravn HP, (Eds): *Ecology and management of giant hogweed (Heracleum mantegazzianum).* Wallingford (UK): CABI; 2007.

104. Reinhardt F, Herle M, Bastiansen F, Streit B: *Ökonomische Folgen der Ausbreitung von gebietsfremden Organismen in Deutschland.* Berlin: Umweltbundesamt; 2003.

105. Andreu J, Vilà M, Hulme PE: An assessment of stakeholder perceptions and management of noxious alien plants in Spain. *Environ Manage* 2009, 43:1244-1255.

106. Dudgeon D, Arthington AH, Gessner MO, Kawabata ZI, Knowler DJ, Lévêque C, Naiman RJ, Prieur-Richard AH, Soto D, Stiassny MLJ, Sullivan CA: Freshwater biodiversity: importance, threats, status and conservation challenges. *Biol Rev* 2006, 81:163-182.

107. Geist J: Integrative freshwater ecology and biodiversity conservation. *Ecol Indic* 2011, in press.

108. Souty-Grosset C, Holdich DM, Noel PY, Reynolds JD, Haffner P, (Eds): *Atlas of crayfish in Europe*. Paris: Museum National d'Histoire Naturelle; 2006. Patrimonines Naturels 64

109. Söderbäck B: Replacement of the native crayfish *Astacus astacus* by the introduced species *Pacifastacus leniusculus* in a Swedish lake: possible causes and mechanisms. *Freshwater Biol* 1995, 33:291-304.

110. Söderhäll K, Cerenius L: The crayfish plague fungus: history and recent advances. *Freshwater Crayfish* 1999, 12:11-35.

111. Rahel FJ: Homogenization of freshwater faunas. *Ann Rev Ecol Syst* 2002, 33:291-315.

112. Longepierre S, Robert A, Levi F, Francour P: How an invasive alga species (*Caulerpa taxifolia*) induces changes in foraging strategies of the benthivorous fish *Mullus surmuletus* in coastal Mediterranean ecosystems. *Biodivers Conserv* 2005, 14:365-376.

113. Lodge DM, Williams S, MacIsaac H, Hayes K, Leung B, Loope L, Reichard S, Mack RN, Moyle PB, Smith M, Andow DA, Carlton JT, McMichael A: Biological invasions: recommendations for policy and management. *Ecol Appl* 2006, 16:2035-2054.

114. Keller RP, Lodge DM, Lewis M, Shogren J: *Bioeconomics of Invasive Species: Integrating Ecology, Economics, Policy and Management*. New York: Oxford University Press; 2009.

115. Molnar JL, Gamboa RL, Revenga C, Spalding MD: Assessing the global threat of invasive species to marine biodiversity. *Front Ecol Environ* 2008, 6:485-492.

116. Ricciardi A: Patterns of invasion in the Laurentian Great Lakes in relation to changes in vector activity. *Divers Distrib* 2006, 12:425-433.

117. Carlton J: Predictions of the arrival of the zebra mussel in North America. *Dreissena Polymorpha Information Review* 1991, 2:1.

118. Finnoff D, Shogren JF, Leung B, Lodge D: Take a risk: preferring prevention over control of biological invaders. *Ecol Econ* 2007, 62:216-222.

119. IUCN: IUCN Guidelines for the Prevention of Biodiversity Loss Caused by Alien Invasive Species. [http://iucn.org] *webcite* Gland (Switzerland): International Union for the Conservation of Nature; 2000.

120. Keller RP, Lodge DM, Finnoff DC: Risk assessment for invasive species produces net bioeconomic benefits. *P Natl Acad Sci USA* 2007, 104:203-207.

121. Miller C, Kettunen M, Shine C: *Scope options for EU action on invasive alien species (IAS)*.Final report for the European Commission. Institute for European Environmental Policy (IEEP), Brussels, Belgium; 2006.

122. Leung B, Lodge DM, Finnoff D, Shogren JF, Lewis MA, Lamberti G: An ounce of prevention or a pound of cure: bioeconomic risk analysis of invasive species. *P Roy Soc B-Biol Sci* 2002, 269:2407-2413.

123. Keller RP, Frang K, Lodge DM: Preventing the spread of invasive species: intervention guided by ecological predictions leads to economic benefits. *Conserv Biol* 2008, 22:80-88.

124. Shine C, Kettunen M, Genovesi P, Essl F, Gollasch S, Rabitsch W, Scalera R, Starfinger U, ten Brink P: *Assessment to support continued development of the EU Strategy to combat invasive alien species. Final Report for the European Commission*. Institute for European Environmental Policy (IEEP), Brussels, Belgium; 2010.

125. Hulme PE, Pyšek P, Nentwig W, Vilà M: Will threat of biological invasions unite the European Union? *Science* 2009, 324:40-41.

126. Essl F, Lambdon P: Alien bryophytes and lichens of Europe. In *Handbook of Alien Species in Europe*. Edited by DAISIE. Dordrecht: Springer; 2009:29-41.

127. Gozlan E, Andreou D, Asaeda T, Beyer K, Bouhadad R, Burnard D, Caiola N: Pan-continental invasion of *Pseudorasbora parva*: towards a better understanding of freshwater fish invasions. *Fish Fish* 2010, 11:315-340.

128. Bernauer D, Jansen W: Recent invasions of alien macroinvertebrates and loss of native species in the upper Rhine river, Germany. *Aquat Invasions* 2006, 1:55-71.

129. Van der Velde G, Rajagopal S, Kelleher B: Ecological impact of crustacean invaders: general considerations and examples from the Rhine river. *Crustacean Iss* 2000, 12:3-33.

130. Pöckl M: Success of the invasive Ponto-Caspian amphipod *Dikerogammarus villosus*by life history traits and reproductive capacity. *Biol Invasions* 2009, 11:2021-2041.

131. Bij de Vaate A, Jazdzewski K: Geographical patterns in range extension of Ponto-Caspian macroinvertebrate species in Europe. *Can J Fish Aquat Sci* 2002, 59:1159-1174.

132. Van derVelde G, Platvoet D: Quagga mussels *Dreissena rostiformis* bugensis (Andrusov, 1897) in the Main River (Germany). *Aquat Invasions* 2007, 2:261-264.

133. Figuerola J, Green AJ: Dispersal of aquatic organisms by waterbirds: a review of past research and priorities for future studies. *Freshwater Biol* 2002, 47:483-494.

134. Hussner A, Van de Weyer K, Gross EM, Hilt S: *Comments on increasing number and abundance of non-indigenous aquatic macrophyte species in Germany*. 2010, 50:519-526. Weed Res

Chapter 6

Pastoralism in Soqotra: External Entanglements and Communal Mutations

Serge D Elie

Yemen Center for Studies and Research, Sanaa, P.O. Box 7305, Yemen

ABSTRACT

This article presents the first study of pastoralism on Soqotra Island, which is the main island of the Republic of Yemen's Soqotra Archipelago in the Indian Ocean. Its inhabitants constitute a once predominantly pastoralist community with a unique language and of mixed ethnicity who are now being enlisted in a state-sponsored and internationally-assisted conservation-with-development experiment. The article seeks to remedy Soqotrans' disenfranchisement from the academic literature on pastoralism around the world. Accordingly, it gives an overview of the nature of pastoralism on the island. This is done through the following analytical tasks: First, it provides a brief geo-cultural contextualization of the island. Second, it describes the underpinnings of a mixed subsistence pastoral regime in terms of livelihood practices, spatial mobility and residential modality. Third, it offers a provisional inventory of the pastoral flocks and herds. Fourth, and finally, the article summarizes the pastoralists' sociocultural changes resulting from Soqotra's entanglement with an externally initiated process of communal transition.

INTRODUCTION: SITUATING SOQOTRA

The Soqotra Archipelago is the Republic of Yemen's ultimate frontier in the Indian Ocean, as it is situated within 250 km of the Somali coast on the African continent, and separated by a distance of about 380 km from the southeastern coast of mainland Yemen on the Arabian Peninsula. The Archipelago is composed of four islands: Soqotra, Abd al-Kuri, Samha and Darsa, in addition to a few rock formations. The Archipelago's inhabitants represent a still vibrant ethno-linguistically indigenous minority community of Southern Arabia as speakers of Soqoṭri, which is one of the six non-Arabic Semitic pre-literate languages of South Arabia that are collectively known as the Modern South Arabian (MSA) languages (see Elie [2012a]).

Soqotra is the largest island in the Arab world (e.g. it is nearly five times the size of the state of Bahrain). Pastoralism, as a non-nomadic transhumant subsistence livelihood with herds of goats, sheep, cows and camels, remains a sociocultural pillar of the rural sector but of diminishing economic importance. Soqotrans, however, have always pursued a multiplicity of subsistence livelihoods such as fishing, date farming and geographically limited vegetable gardening. The practice of these livelihoods is contingent on seasons, variations in agro-ecological zones, and further constrained by the five-month summer monsoon, during which economic life comes to a relative standstill. These subsistence activities are being supplemented by increasing employment opportunities through an emerging modern economic sector (e.g. construction, tourism, salaried posts in the island's government bureaucracy as clerks, police and military personnel), and further complemented by remittances from the Soqotran *émigré* community in the Arabian Gulf.

Soqotra is currently being described as the 'Galapagos of the Indian Ocean' because of its endemic biodiversity, which is ranked into fourth place among the ten richest of the world's oceanic islands that are distinguished by the richness of their biodiversity. This led to the rediscovery of the island at the dawn of the twenty-first century with its designation in 1996 by the Yemeni government as a biodiversity preserve. In turn, this has engendered a dual incorporation process of the island and its people: on the one hand, the Yemeni government's modernization of its infrastructure and consolidation of its political incorporation into the national community; and, on the other hand, a United Nations-led internationalization of its economy through the implementation of an environmental protection and ecotourism development programme. This state-sponsored and internationally assisted dual incorporation has spawned an internal social transformation and communal transition process. The effects of the latter on pastoralism in Soqotra are the focus of this article.

STUDY AREA

Soqotra Island, which is the primary focus of this study as the main island of the Soqotra Archipelago, encompasses an area of 3,625 km^2 and measures 133 km (82.6 miles) from west to east and 43 km (26.7 miles) from north to south. Its population of approximately 50,000 is divided into a few hundred clans each composed of a few extended families organized into paternal or fraternal joint households that never consolidated into tribal formations and are occupying over 600 hamlets that are dispersed throughout the island. The clans' residence in these hamlets is contingent upon seasons, weather condition and availability of resources (e.g. water and pasture).

The dominant topographical ordering of Soqotra's landscape is the natural succession of certain topographic features over most of the island's landmass that constitute five physiographic areas with varied types of vegetation: coastal arid plains (*halla* in Soqoṭri), followed by high granite mountains (*fedenhen*), and elevated limestone plateaus (*dimidi* or *igalis*), separated in places by internal alluvial valleys (*shi'ib*) or wadis (ephemeral seasonal river beds), and interior semi-arid plains (*sheteh*) that are punctuated in places with rolling hills, and finally sand dune-filled coastal plains again. This undulating topography harmonizes with, indeed dictates, the Soqotrans' practice of a seasonal altitudinal transhumance form of pastoralism that alternates with the two monsoon seasons on the island: the North East (winter) monsoon from November to March, and the South West (summer) monsoon from May to September. The island lies on the margin of the sub-equatorial and northern tropical climate belts. Accordingly, mean annual rainfall is 216 mm which varies across the island according to exposure and elevation: on the coastal plains, the annual precipitation may be as little as 125 mm or even absent altogether; while in the mountains, it can reach exceptional levels of 1,000 mm (De Sanctis et al. [2013]: 149). In a context of high variability in rainfall and in key resource availability (e.g. water and pastures), the islanders demarcate Soqotra's landscape into zones that do not merely designate geographical locations but represent a series of natural frontiers made up of topographic signposts, climatic and altitudinal variations, agro-ecological particularities, and the types of subsistence activities they afford (see Elie [2006]: chapter 3) (Figure 1).

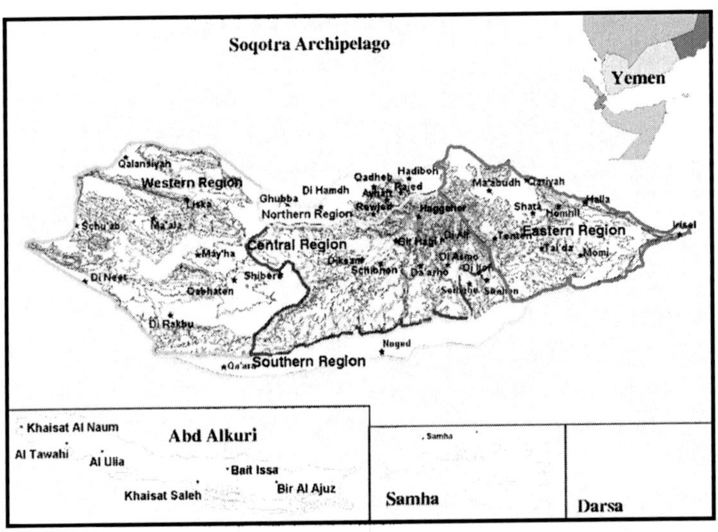

Figure 1: Map of Soqotra.

METHODS

The study that follows is based on long-term anthropological fieldwork starting in February 2002 for 18 months, and subsequently complemented by multiple visits over a ten-year period up to January 2014. My fieldwork method deployed a recursive movement between multiple locations on the island in the form of 'sites hopping', which refers to the practice of the short stay, and a pattern of commuting at irregular intervals to locations all over Soqotra in a manner that simulated the Soqotrans' opportunistic spatial displacements in pursuit of a multiplicity of livelihood-sustaining activities. This approach to fieldwork was most appropriate to the topography of life on the island, which was characterized by a population distributed into hundreds of hamlets of seasonal occupation, each with a small number of inhabitants. This context undermined the viability of the traditional one-village ethnography; instead, it obliged recourse to a research approach that privileged a cross-regional focus on the socio-spatial configuration of Soqotran pastoralists' lifeworld (see Elie [2012b]). This study of a formerly predominantly pastoral polity within a sub-national territorial entity of the Yemeni state that is presently entangled in a comprehensive communal transformation process emphasizes the changing interactional dynamics between Soqotran pastoralists' ecological praxes, environmental milieu and the pastoral herds. Accordingly, it focuses on the transitional processes engendered on the island and their impacts on Soqotra's pastoralists and their way of life.

Due to space limitations, the article provides a descriptive and analytical overview of only some of the key aspects of pastoralism in Soqotra. Missing here, among other aspects, are the contemporary workings of the pastoral economy in terms of the processes of production, distribution and consumption, which will be presented in a subsequent paper (Elie [2014]).

This article is organized as follows: First, it describes the specificities of Soqotran pastoralism in terms of (a) a mixed mode of subsistence driven by environmental opportunism, (b) the nature and seasonality of the spatial mobility of Soqotran pastoralists, and (c) the geography of residence in terms of the distribution of clans and villages throughout the island's sub-regions. Second, it provides a provisional inventory of the animal population. And third, it concludes with a summary analysis of the externally engendered transitional processes on the island and the resulting mutations in its pastoral way of life.

LIVELIHOOD PRACTICES: PILLARS OF A MIXED SUBSISTENCE STRATEGY

This section describes the three pivotal aspects of pastoralism in Soqotra: the environmentally and historically contingent valuation of livelihood practices, the vectors of Soqotran pastoralists' spatial mobility, and the economic geography of their pattern of residence. Soqotrans' subsistence pastoralism is practiced according to the rhythm of the two monsoons, and the seasonally programmed availability, or scarcity, of key environmental resources. Prior to the end of the Sultanate in 1967, hunger was a seasonal fatality as the environmental dependency of the islanders was at its apogee. This dependency was relieved only intermittently, and not always reliably, by the traffic of boats carrying *maqdereh* (maize) from East Africa in exchange for the locally produced butter oil. Hence, every nook and cranny of the available territory was explored in search of environmental resources. Moreover, this quest for resources was not only for pastures and water – although these were critical to ensuring the survival of their animal herds – but also for wild comestible items (e.g. fruits, plants, birds, honey) to supplement the main diet made up of meat (rather infrequently), milk, dates, home-grown millet and imported maize, which was subsequently replaced by rice in the 1970s. Equally important was the search for items that could be bartered: for example, collecting fruits (such as tamarind), aloe juice, lichens from limestone mountains, resin from the bark of various trees, weaving of sheep wool rugs, gathering timber and vegetable gardening. What was significant about this quest for resources was that it occurred in a context in which survival was a permanent challenge, and where livelihood opportunities have always varied according to seasons. Hence, subsistence diversity was a prerequisite and environmental opportunism (i.e. maximizing the use of all available natural resources) the obligatory means of ensuring survival. In such a context, it becomes clear why it is erroneous to assume that 'ostensibly predominant subsistence techniques represent total subsistence strategies' (Ellen [1982]: 170).

It is in this light that two recent attempts at characterizing the Soqotrans' mode of subsistence are to be evaluated. First, Naumkin

([1993]) insisted on dichotomizing Soqotrans into two 'economic-cultural types': the mountain-dwelling pastoralist, and the coastal fishermen. Accordingly, he argued against the designation of a third 'economic-cultural' category among Soqotrans as agropastoralist, because (a) of the minimal proportion of time allocated to land cultivation, (b) the insignificant quantity of palm trees owned, (c) the limited knowledge of other forms of agriculture, and (d) more importantly that the value of land is in its pastures not in the date palm trees. Naumkin's argument was partly motivated by a theoretical exercise that sought to reject the applicability of the dominant view that agriculture preceded pastoralism. The second attempt is by the European Union-funded team that prepared the *Socotra Archipelago Master Plan* (SAMP), which characterized Soqotrans as forming a holistic community of pastoralists. As they explained: 'For this report Socotrans are considered to form a single pastoral community... [with] an economic system in which the herding of domestic ruminant animals on open rangeland is the dominant economic activity' (Commission of the European Communities EU [2000a]: 4–1). This undifferentiated characterization of the Soqotran population was motivated by the need to identify the local agent of an environment-led development policy, and the targeting of the Soqotran pastoralist for re-invention as an intuitive eco-conservationist. In both instances, the authors were aware that this pastoral economic system had undergone a mutation since the demise of the Sultanate, and thus, the dichotomous or holistic characterization amounted to an expedient reductionism: the first to validate a theory, and the second to justify a policy.

However, recourse to either discrete or amalgamating descriptive categories is not the only available analytical strategy. More appropriate is to look at Soqotran pastoralism as a mode of subsistence encompassing 'the aggregate of extractive processes characterizing a particular people' (Ellen [1994]: 198). In other words, a mode of subsistence is constituted by an ensemble of 'subsistence techniques' that are the means with which resources are appropriated from the environment. The determinant of when a technique or a combination of them is used is what I call variable geography of livelihood opportunities, which refers to the different sets of subsistence activities afforded by the island's different geographical regions and their particular resource endowments. It is this ineluctable recourse to multiple means of subsistence that characterizes Soqotrans' livelihood making. It is a situation that cannot be analytically elucidated by specifying pure type of economic agents or generalizing about an undifferentiated pastoral community. Furthermore, the multiple livelihood practices of Soqotrans, and thus their lack of livelihood specialization, problematize the standard definition of who are pastoralists: 'populations for which livestock-related activities are the sources of more than half of total household income' (Swift and Toulmin [1992]).

Core vs. Auxiliary livelihood: a contingently Alternating subsistence Repertoire

The above situation entails the analytical challenge of articulating this environmental opportunism-inducing geographical context – that is, each region allows a different pairing of subsistence techniques and thus multiple shifts in subsistence emphasis – with the assertion that pastoralism is the dominant subsistence strategy of the Soqotran community. This articulation can be effectuated through the adoption of the distinction between core (i.e. central, foundational) and auxiliary (i.e., supplementary) subsistence activities. However, these two categories are not stable, as they alternate along three dimensions: environmental context, gendered division of household labour and historical period. In the following discussion, I illustrate how core and auxiliary subsistence activities vary along these three dimensions subject to the seasonal calendar:

First, a subsistence activity is either core or auxiliary depending on the physiographic areas (i.e. coastal plains, high mountains, high plateaux, interior valleys and interior plains), their resource endowments, climactic condition and access to alternative livelihood opportunities outside the pastoralists' primary residential domain. I schematically illustrate the alternation between subsistence activities by comparing mostly herding and fishing while neglecting a miscellany of other activities: On the coastal plains, fishing is the core activity, while date cultivation and herding are auxiliary ones. Also, for a proportion of the coastal population, especially in the north, salaried employment is the core activity with trading as an auxiliary activity, and where the cultivation of dates is an un-remunerative tradition-bound endeavour, and fishing is either an occasional leisure, or supplementary income, activity. In the high mountains, herding is the core activity with negligible participation in auxiliary activities such as date cultivation. In the high plateau areas, herding is the core activity supplemented, in the central region, by date cultivation and the collection of resin from incense-bearing trees, and in the western region, some pastoralists migrate toward the southwestern coast for seasonal fishing. In the internal valleys, herding is the core activity with date cultivation as an auxiliary one, but assumes core status seasonally. In the internal plains, herding is the core activity, which is seasonally complemented by fishing on the northern coast as the auxiliary activity. However, the subsistence activities, whether core or auxiliary, in all the island's areas are regulated by the seasonal calendar, and in particular by the occurrence of monsoon winds and seasonal rains.

Second, a gendered division of household labour further distinguishes core and auxiliary activities. In general, core subsistence activities are the responsibility of men (e.g. herding[a], milking, date farming and fishing), while the auxiliary ones belong to women (e.g. all household chores: water and wood collecting, cooking, milk churning, horticulture, rug weaving, pottery making). However, this gendered division concerning the performance of core vs. auxiliary chores, alternated in response to historically induced changes in the rural economy. For example, herding was principally, but not exclusively, a male activity as well as milking, but the processing of milk into different dairy by-products, but most importantly butter oil, which was the pillar of the pastoral economy, was a female activity. With the discontinuation of the export of butter oil that followed the demise of the Sultanate, and the increasing importance of fishing encouraged by the succeeding Socialist administration in the 1970s, herding, and even the milking of animals, at least in some places and during the extended absence of men (as milking remains taboo for women[b] in certain parts, as in the central mountain region in contrast to the southern coastal plains), and milk churning for the preparation of *hilub* (butter milk) was taken up by men, at least in some places. I witnessed one instance of this in the village of Aduno in the central internal valleys. This reallocation of task within the household was due to the seasonal migration of men to the coast for fishing and to their frequent search for odd jobs in response to the increasingly cash exigent economy. In sum, this is the extent of the division of labour in Soqotra's pastoral economy, as there is no pastoralist-agriculturalist or nomad-sedentary occupational dichotomy and the corollary social antagonism that is supposed to animate the political dynamics of pastoral societies.

[a]The herding of sheep seems to be the primary, if not exclusive, responsibility of women; however, males in the western part of the island herded sheep as well. More generally, men were in charge of adult livestock, while women were in charge of immature ones (personal communication with Dr. Miranda Morris).

[b]In this regard, Soqotra seems to be an exception, given that 'A constant factor among pastoral populations is the assignation of milk and milking to women' (Blench [2000]: 30). In Soqotra, this is true only in the case of sheep, as noted above.

Third, and lastly, whether subsistence activities were core or auxiliary was a result of a historical process driven by episodes of political incorporation and their economic entailments. Three such entailments can be identified: (1) the reconfiguration of the nature of subsistence activities in terms of a gradual shift toward alternative income sources, (2) the introduction of new consumption items and a corresponding reprioritization of food resources, and (3) the reorientation of the local economy from an environment-based and barter-mediated to a service-based and cash-driven one. The initial period of the still-in-progress historical mutation of the island's pastoral economy can be dated from the demise of the Sultanate and the beginning of the Socialist administration on the island in 1967. Soqotra can still be designated as a pastoralist community, but only if pastoralism is understood as a defining cultural leitmotif, not as the dominant economic activity. This observation does not lessen the importance of pastoralism in the Soqotrans' repertoire of socially valued activities, but seeks to distinguish between pastoralism's cultural function and its economic contribution. In terms of its cultural function, herding is valued for the social anchor – a sense of existential rootedness to the land – that it confers, rather than for its economic value. In terms of its economic contribution, herding is primarily a subsidiary food production activity (i.e. milk), since the demise of the inter-continental market for butter oil was not replaced by a market for meat. Soqotran pastoralists have become increasingly reliant on odd jobs to make ends meet, thus generating a situation in which pastoralists have moved 'from living off, to living for' their herds.

It follows from the above that the pastoral mode of subsistence in Soqotra has always encompassed multiple means of drawing resources from the environment. Furthermore, these means have varied according to agro-ecological zones, intra-household shifts in labour allocation and historically induced economic changes. These susceptibilities are intrinsic to the pastoral mode of subsistence and are generative of a chronic sense of livelihood insecurity. This raises the issue of pastoralism's sustainability as a way of life, which is further threatened by increasingly 'strong circumstantial evidence that the islands have been drying out over the last few hundred years' (Miller et al. [2003]: 79) and the ensuing chronic scarcity of resource availability.

Spatial Mobility: Seasonal Peripatetics

The nature of pastoral movements – i.e. frequency, directionality, timing, fixed or variable destination, etc. – is the key defining feature of pastoralism, hence the need to specify the type of movements of Soqotran pastoralists. The term transhumance has been used to characterize the movement pattern of Soqotran pastoralists (Naumkin [1993]; Commission of the European Communities EU [2000a]; Morris [2002]). The use of this term is justified by the island's undulating topography and the fact that it was first used to describe seasonal herd movements that alternated between mountains and lowland pastures: 'During one part of the year the livestock is kept in mountain pastures and during the other parts is driven to lower zones'. The main purpose was to 'use other areas as seasonal pastures when they are at their optimum productivity' (Khazanov[1994]: 23). Accordingly, transhumance is a spatially limited and seasonally alternating pattern of movement by (some of) the pastoralists and their herds from permanent village bases located in mountains to seasonal encampments in lowland pastures (or vice versa), due to altitudinal and seasonal variations in pasture availability. Moreover, transhumance proper can co-exist, and has done so, with other forms of pastoral movement, and thus the need to identify them accordingly. This is the approach that informs my description of Soqotran pastoralists' movements. Soqotrans seasonally reincarnate into practitioners of different livelihoods: pastoralist, fisherman, date palm tree cultivator and odd-jobber. This self-fashioning into occupational peripatetics is due to the seasonal constraints imposed on the kinds of livelihood that can be pursued at different periods of the year. These seasonal constraints provide the impetus to the different kinds of movement undertaken by Soqotrans, which I describe below.

Transhumance: Intra-Territorial Migration

This is the principal movement among the spatial displacement patterns followed by Soqotrans, and it is the defining feature of Soqotran pastoralism. Transhumance was the foundational human practice, as a mode of livelihood, on the island. It entails a regular long-distance migration to a known destination and return to an established home base. It determines land use practices as well as settlement patterns across the island, as it entails a two-part movement across eco-topographic zones, and geographical regions, which is synchronized with the two monsoon seasons. The first movement is *mezhīro* from higher to lowland pastures and occurs toward the end of the SW monsoon season of *horf* in early October. The return movement is *merqīyo* from lower to higher grazing grounds, and it takes place toward the end of the NE monsoon season of*sereb* in January or after the cold weather has subsided at the higher elevation. In effect, it is a movement from a home base, an area of permanent residence to an area of temporary residence and back. The distinctive aspect of this movement is that it takes place primarily within the territorial domains of the clan collective. However, in some cases, it involves displacement to the territory of another clan in another geographical region (e.g. from central mountains to eastern plains) based on long-standing agreements. This movement is primarily for the sake of the animals. It is ultimately dependent on the seasonal variation in the weather, on whether rains have occurred, and thus on the availability of pastures and water in the places of destination, and on the condition of pastures at home base. Given the changing pattern of the weather and its impact on environmental resources, this movement's timing tends to fluctuate. Thus, it is a combination of push factors from the home base and pull factors toward the destination. The main strategies, or mobility rationales, underpinning the Soqotrans' practice of transhumance are as follows:

- Resource conservation: this entails the search for alternative pastures, as a means of preventing overgrazing (*māṣel*), and to allow the land to lie fallow for its regeneration, which Soqotran pastoralists call *kor inābut*: 'so that the land can revive' (Morris [2002]: 473).
- Weather avoidance: in areas where it gets very cold, such as in the high mountains (e.g. Hajhir Mountains) and high plateau

areas, pastoralists migrate to the shelter of lowland areas.

- Follow the rain: this entails moving from dry to wet areas in search of water and pastures when rains have failed at home base.
- Herd 're-socialization': this is especially for the goats in mountainous areas or in valleys with dense vegetation cover where they divagate unattended, and must be taken toward elevated plateau areas with sparse vegetation where they can be controlled, and thus prevented from lapsing into a feral state.
- Livelihood alternation: pastoralists – especially those who are located in the high mountain and plateau areas and have access to date palm tree gardens in the valleys below – move to their gardens to harvest dates, while simultaneously avoiding the winds of the SW monsoon, which coincide with the harvesting period.
- Maintenance of traditional access to land: this is specific to groups of pastoralists who transhume to land owned by others, and the use of which is based on unwritten agreements between elders of generations ago in the different communities concerned and that have sedimented over the years into an irrevocable tradition. However, these agreements are becoming burdensome for the receiving community (e.g. between pastoralist communities of the central mountains and the eastern plains); hence, movement is partly about ensuring continued access to such land.

Situational Nomadism: Cross-Territorial Displacement

The recurrent irregularity of the rain seasons, and the deleterious effects on pastures and water availability, has given rise to a displacement pattern that lacks the regularity, directionality or the altitudinal characteristics of the *mezhīro-merqīyo* movement. This situationally constrained nomadism is a movement that is motivated by a crisis situation in which water and pastures have been exhausted by a prolong absence of rain; thus, it might be best understood as a drought management strategy. The main manifestation of this form of occasional nomadism is the *mat'ino*: this movement occurs not only outside the regular transhumant pattern but also more importantly outside the territorial domain of a given clan collective in search of 'drought retreats'. This movement is a form of induced nomadism as the ultimate destination is unknown because it depends on where water and pastures are found; hence, movement becomes erratic, and is no longer restricted to the regular vertical (up-down) motion, but entails a horizontal (east–west) movement pattern as well. Depending on the extent of the geographical space affected by the dry spell, the concerned clans might have to travel very far beyond their region to find adequate pastures. This movement takes place usually during the transitional seasons of *gyahsh*, *doti* or*serebhen*, which is a period of high livelihood stress. A scouting mission to adjoining areas to observe and inquire about the availability of water and pastures precedes the move. If these two resources are found in sufficient quantity, then a formal request is made to the responsible persons (*muqaddam*) of the receiving village or area. This request (*yish'hallin*) for permission to move into the territory of the clan is made for a specific period of time – usually until it has rained in the villagers' land – to graze animals and set up household. This request might involve an individual family or a whole group. Once permission is granted, the next step is *ta'an*, which involves moving the herd and setting up household under strict arrangements regarding what land space to be used and the house to be occupied by the family. Depending on prior agreement with the landowners, a compensation fee (*mizbido* or *gizire* of *kire* according to region) usually in kind could be paid to them. This movement represents another form of mutual aid practices, as part of culturally elaborated conventions for intra-communal socio-environmental relations.

Agropastoral Movement: Food Harvesting Relocation

This movement not only illustrates the notion of seasonal peripatetics, as it involves the mutation of coastal fishermen and hinterland pastoralists into date farmers, but also the seasonal as well as historical alternation between core and auxiliary subsistence activity. Agropastoral movement concerns essentially date harvesting in the regions where there are date gardens. In historical terms, date harvesting was the high point in the cycle of livelihood making, as it brought relief to the seasonal hunger that befell Soqotrans just prior to, and during, the *horf* season, as dates were considered the most effective hunger suppression food. The date garden was a wealth and social status marker and was the main form of investment from the Soqotran diaspora in the Arabian Gulf until the early 1990s. The movement to date gardens is called *yaharuf* from the verb *haruf*meaning 'to harvest dates'. This movement took place just prior to the beginning of the SW monsoon season of *horf* from the fishermen or the pastoralists' permanent residence to the date gardens during the harvesting period, which is during the months of June and July. In some places, as in the central region (e.g. Hajhir Dit Fahed), the people use a relay system, as they go to the date gardens for a few days and return to their village for a period of rest. In other places, such as Hadiboh, date farmers move to their date gardens for the duration of the harvest, at which time Hadiboh is partially emptied of its residents. This movement is still practiced by many people, although it is less extensive than during the Sultanate period, when it constituted a major seasonal event on the island (see Elie [2004]: 73). This movement is partly linked to the end of the school year and the beginning of a three-month vacation period, and entails the seasonal desertion of some settlements on the northern coast either toward date gardens, or more generally for hinterland settlements in order to escape the winds.

Finally, movement in all of its forms remains an intrinsically weak link in the pastoral mode of livelihood, as it could undermine commitment to it. As Blench ([2000]) explains, 'Pastoralists exist along a gradient of willingness and capacity to move'. It is this 'willingness' , which can be gradually compromised by the availability of alternative means of livelihood that are less demanding than pastoralism and less susceptible to seasonal limitations or uncertainties. This threshold, however, has yet to be crossed, at least not permanently – it is straddled instead – as no livelihood alternative has yet to escape the ineluctable seasonal constraints on all economic activities on the island. The only exception, thus far, is the limited number of salaried jobs available through government posts and as staff member of the few projects of the UN and international NGOs. Indeed, it is the exigencies of the seasonal calendar that determine the movements associated with all livelihood activities on the island. Accordingly, the seasonal calendar presented in Table 1 seeks to articulate the interaction between seasons, climactic conditions and their socio-economic ramifications for life on Soqotra.

Table 1: Soqotra's seasonal calendar

Season	Period	Climate	Socio-economic activities
Qeyat(3–4)	Transitional post-winter monsoon season: from March to April.	Rains are sporadic and spotty, but mostly rare. Water and grazing availability is limited.	Pollination of date-palm trees wherever they are cultivated.
	Occurs in the spring, but considered as Soqotran 'summer'.	Weather is warm to pleasant. Westerly winds affect the southern part of the island.	Celebration of Muslim holiday *d al- dh* , the main occasion for the sale of large numbers of livestock, esp. sheep, male goats (*mi'shur*) and ram (*kub*) for obligatory sacrificial ritual. Its seasonal occurrence varies as it depends on the lunar calendar.
Gyahsh(4–5)	Spring transitional period between the two monsoons: from mid-April to mid-May.	Expected to bring pre-summer rains to some parts of the island.	Out of season semi-nomadic movement in search of appropriate pastures either in or out of territorial domains.
	Overlaps the end of spring and beginning of summer, and buffers the dry *doti* season.	Relatively hot, as SW winds die down and before SE winds start.	End of tourism season.
			Import of animal feed – barley (*sha'ir*), flour (*qahm*) and wheat shaft (*sabus*) – in expectation of absence of rains in the *horf*season.
Doti (4–6)	Transitional pre-summer monsoon season: from late April to early June.	Unpredictable summer rains, mostly in highland areas, are seen as harbinger of whether or not drought will follow.	Livestock come into season, and sheep and goats are mated to give birth in *sereb*. Prior to this period, breeding would normally be discouraged in order to avoid birth during water and grazing scarce*horf* season.
	Straddles late spring and early summer.	Weather is generally cool and breezy if it rains.	Out of season semi-nomadic movement from dry to wet areas to feed animals prior to the long dry summer monsoon season.
		Grazing grounds in wet areas are covered with grasses. Water availability is variable.	Preferred marriage season, as sufficient cash was accumulated during the previous *sereb* season.
			Extensive out-migration of fishermen from coastal villages around the island and of pastoralists from the hinterland to villages in Hadhramawt and Mahra for one to three months to work as fishermen, herders, or construction labourers.

Horf (6–9)	SW summer monsoon season is the longest on the island. Lasts nearly five months: from mid-June to end of September or early October.	Gale force *medeh* winds averaging 70 mph. Strongest impact on northern Coastal Plains (CP), and High Mountains (HM), and mildest on Internal Valleys (IV), and Internal Plains (IP).	Economically moribund time of year, as sea is closed to fishing and shipping. Some fishermen from the northern and southern coasts migrate to the mainland to work as fishermen, among other temporary occupations.
	Straddles end of summer and mid-fall.	Hottest period of the year and very dry. Absence of rains, thus scarcity of grazing and water.	Closure of school for summer vacation (June-August) and partial migration of residents from northern coast toward hinterland settlements and to date gardens.
		Depleting rain water catchments (*karifs* esp. in west and east) are replenished with purchased water trucked in from Hadiboh wells.	Season for date harvesting, drying and storing for later consumption (esp. during Ramadhan). Also, harvesting of aloe juice (esp. in the west).
			Main season for honey production by hundreds of bee-keepers in villages around the island, which has partially replaced wild honey harvesting.
			Departure of mainland workers and stoppage of all construction works and road building.
			Washing and shearing of sheep wool for weaving rugs, especially in the east where rug production is integral to household activity.
			Onset of *mezhi* period when *medeh* winds die down and signals the beginning of the main transhumance movement, *mezh ro*: from higher to lower grounds.

Serebhen(9–10)	Autumn transitional period that straddles end of *horf* and beginning of *sereb*. Critical period for pastoral sts.	Winter rains late in coming, extending further the preceding five-month dry season.	This period serves as catalyst of the *mezh ro* transhumant movement, which is usually within the territorial domain of the clan collectives. The severity of water scarcity might necessitate resorting to alternative coping strategies:
	From late September and mid-October: transitional months between summer and winter.	Water levels at their lowest, and rangelands are bare, esp. in CP, IP and HP areas.	(a) Activation of mutual aid system, as most affected pastoralists canvass other territories seeking permission from owners to graze their livestock.
			(b) Purchasing of cereals such as barley, wheat shaft and flour to feed the herds.
			(c) Selective cutting of particular trees, based on local knowledge repertoire, to use as fodder in the absence of grass.
			Reopening of schools (Sept-May) and return migration from the hinterland to Hadiboh town especially, but to all coastal settlements.

Sereb(10–2)	NE winter monsoon season: from late October to early February.	Main rainy season from October to December.	Reactivation of economic life, as sea opens for shipping and fishing.
	Straddles the fall and winter.	Replenishment of pastures.	Ecotourism season begins.
		High water availability in most regions and the replenishment of rain water catchments (*karifs*and *lims*).	Parturition season (esp. for goats and sheep), and animals are lactating. Main period for pastoralists to sell two-week-old male kids in Hadiboh.
			Main season for milk production and consumption, which reduces dependence on some purchased food items. Also, production of butter oil from cow and goat milk for sale in 1-litre plastic bottles.
			Breeding season for cattle in central and eastern regions.
		Milder *serbihi* winds, cold in higher elevation and cool/pleasant in lowland areas.	Home gardening activities (planting and caring) start early in the season.
			Greater availability for odd-jobbing opportunities (e.g. construction, tourist guides).
			Preferred visiting period of Soqotran *émigrés* from the Gulf.
			Migration from the hinterland to the coast for fishing.
			Second honey collecting season depending on rains.
			Buying and storing of food (esp. sacks of rice and flour), as form of food security, to last through the next *horf* season, when there is usually a shortage in local supply or a substantial price hike.
			Ramadhan, whatever the season of its occurrence, is a period of custom-obligated meat consumption. The occurrence varies depending on the lunar calendar. This is followed by *d al- dh* , which is a major occasion for the local sale of livestock.
			Return of mainland migrant workers.
			Onset of *zuhum* period, the end of *sereb*, when kids and lambs are weaned. And beginning of *meroqihi*, and the return transhumance movement, *merq yo*: from lower to higher grounds. This takes place toward the end of January, or late
			Harvesting of vegetables and fruits FROM HOME gardens around the island, which are primarily for sale (and not for home consumption) in Hadiboh until the end of February.

Note: The table is based on fieldwork observations and complemented by personal communication with Dr. Morris, and from Morris ([2002]), and Miller and Morris ([2004]) especially for the Soqo□ri terms, the timing of the seasons and some of the related socio-economic activities. Under the 'Season' column, numbers are provided in parenthesis, which refer to the months of the year, as Soqotrans use Arabic numbers to designate months – e.g. month one (shahr wahed). Under the 'Period' column, the seasons of spring, summer, fall and winter are referred to as a means of identifying the seasonal equivalence with the solstice calendar.

Elie

Elie Pastoralism 2014 4:16 doi: 10.1186/s13570-014-0016-3

Geography of Residence: Regions, Villages and Clans

Soqotrans' occupation of the island's landscape is characterized by the wide dispersal of settlements. It is as if an 'atomization imperative' was driving them to spread themselves in a myriad of micro-encampments throughout the landscape. This widely dispersed residential pattern, which was partly driven by the unpredictable seasonal availability, and alternating topographic location, of key environmental resources, and further mediated by contingent historical forces, has determined the shifting geographic location of livelihood opportunities. Table 2 attempts to present the current, but still evolving, configuration of the residential geography of Soqotrans. In the table, the island is divided into five regions, which are the actual geo-political zones established for political management purposes – i.e. demarcation of traditional local representative (shaykh) jurisdiction – and encompassing 43 areas. However, the table lists 45 areas; this is due to the inclusion of Hadiboh suburb and Sheteh Qalansiyah (lowlands of Qalansiyah), which are not political jurisdictions but geographical demarcations. These areas are in fact informal sub-districts demarcating the political domain of the 43 locally selected and government approved shaykhs as well as the self-appointed ones. They are informal since they are not recognized in the Local Authority Law enacted in 2000 and are subsidiary to the formally established 36 sub-districts constituting the Eastern (Hadiboh) and Western (Qalansiyah) Districts of the island, and for which Local Councilors are elected (see Elie [2009] for details). The order, in which the areas are presented in the table, is from the perspective of traveling from Hadiboh town, the island's political centre, toward each region. The topographic designations of each area reflect the landscape zoning already discussed above. Each area refers to an agglomeration of villages spread over a given territory and is named after, but not in all instances, the main village in which a mosque (esp. for Friday prayers), a clinic and a school are located, and for which a shaykh is either officially designated or self-appointed. Finally, each area is further sub-divided into individual villages and hamlets shared by clan collectives and headed by a clan or village leader the*muqaddam*.

Table 2: Geography of residence in Soqotra

Regions	Areas	Topography	Villages	Clans	Particularities
Northern 32% (14,000)	Hadiboh	Coast	Town	NA	Heterogeneous clan as well as ethnic composition with approximately 83 clans sharing 103 villages, thus giving it a centripetal settlement pattern. Its share of 32% of the population represents the highest concentration of relatively sedentary inhabitants on the island. They are engaged in diversified occupations (govt. staff, fishermen, agropastoralists, petty traders, etc.). There is herding activity of goats mostly. There is higher access to public amenities, due to its location as the island's political and economic center, main seaport and airport.
	Hadiboh (suburb)	CP	38	NA	
	Qadheb	CP	5	12	
	Ayhaft	HM	8	5	
	Rewjed	HM	23*	6	
	Di Hamdh	CP	12	43	
	Ghubba	CP	17	17	
Central 7% (3,000)	Diksam	HP	111*	9	Highly homogeneous clan composition with highest number of villages (358) per individual clans (67), thus strong centrifugal settlement pattern, which reflects the practice of an extensive pastoralism, and on which the inhabitants are highly dependent. Date cultivation is restricted to the IV areas. Has lowest level of population (7%) with a relatively specialized livelihood, as semi-nomadic pastoralists possessing a diverse herd composition, in which cows are the dominant animal. Has the least access to public amenities.
	Shibhon	HP	30*	3	
	Da'arho	IV	12	1	
	Difshe	IV	11	3	
	Di Asmo	IV	10	2	
	Klim	HP	9	5	
	Hajhir (3 parts)	HM	122*	19	
	Sha'ab Di Alf	IV	23	17	
	Sha'ab Di Ilofi	IV	18	8	

Southern 10% (4,600)	Iglisu Serbehe	HP	9	5	Moderately heterogeneous social composition with relatively proportionate distribution of clans (36) to villages (53), thus a centripetal settlement organization facilitated by the collective nature of primary livelihood activity, fishing (esp. lobster). This is complemented by agropastoral activities (i.e. date and vegetable cultivation, and herding). Ranks fourth in population proportion (10%) with inhabitants of variable mobility, and inadequate access to public amenities.
	Sirahen	HP	10	7	
	Mahfirhen	CP	1	6	
	Noged	CP	11	7	
	Qa'ara	CP	22	11	
Eastern 19% (8,400)	Ma'abudh	IV	12	21	Relatively heterogeneous clan composition with greater number of clans (137) than villages (124), thus having a centripetal settlement pattern. Region straddles multiple topographical terrains, thus population has access to diverse livelihood sources (e.g. fishing, date and vegetable gardening, and herding). This is complemented by (in-kind) remittances from relatives in the Gulf and local government jobs in Hadiboh. Ranks third in population proportion (19%) with moderate availability of public amenities.
	Dibna	CP	4	1	
	Tentehen	IP	16	19	
	Tiyda' (Shizeb)	IP	18	10	
	Sheteh	IP	21	25	
	Momi	HP	74*	24	
	Homhil	HP	14	5	
	Qariyah	CP	4	13	
	Halla	CP	11	13	
	Irisel	CP	10	6	

Western 32% (14,000)	Qabhitin	IP	32	14	Homogeneous clan composition with relatively unbalanced distribution of clans (86) to villages (189), thus a centrifugal settlement pattern. Population is primarily engaged in goat herding and some sheep, as there are no camels, cows or donkeys; highly mobile pastoralists due to high water scarcity given the dominant topography of arid internal plains; minimal to non-existent access to public amenities. Ranks second in population proportion (32%). However, this figure is based on the geographical division of the island, and not on its political division which allocates a sizable portion of the west to the Eastern District (e.g. Qabhitin). Some pastoralists undertake seasonal fishing on the northern and southern coasts. Indeed, the fishing settlements of the western end of the southern plain are made up of people from this region. With the exception of Qalansiyah town and Qeso village, the region is bereft of vegetable gardens and date cultivation.
	Kishin	IP	10	7	
	Di Rakbu	IP	15	6	
	Shibere	IP	10	4	
	Mayha	IP	47*	16	
	Ma'ala	HP	49*	9	
	Shu'ab	CP	3	7	
	Di Neet	CP	6	5	
	Stum	IP	4	7	
	Liska	IP	2	1	
	Sheteh Qalansiyah	IP	11	10	
	Qalansiyah	Coast	Town	NA	
	Samha	Island	NA	NA	
	Abd al-Kuri	Island	6	NA	
Total	45		869	409	

CP, coastal plains; HM, high mountains; HP, high plateau; IV, interior valleys; IP, interior plains.

Elie

Elie Pastoralism 2014 4:16 doi: 10.1186/s13570-014-0016-3

The figures for villages and clans are based on a number of sources: the author's systematic, but not exhaustive, collection of information directly in the regions; from knowledgeable individuals from the regions; and a systematic review of the data from the 2004 census. It is on the basis of the latter that the population distribution of the island was arrived at. The social information is based on my personal observations, complemented by multiple informants. Also, the number of villages identified (869) includes hamlets as well as seasonally uninhabited grazing areas used for transhumance, and are just as important to pastoralists as are inhabited villages. The actual number of settlements might be slightly higher than 900, and the number of clans might be nearer 500. In contrast, the census in 1994 was based on 880 villages, and that in 2004 covered 629 villages. The inclusion of uninhabited grazing areas is most evident in the central and western regions and is identified with an asterisk (*) under the 'Villages' column. The number of clan collectives (409) does not always include clan sub-divisions, which seems widespread (e.g. in the internal plains of the western region). Also, the term 'tribes' is not used in the table, although that would be in keeping with Soqotrans' self-reference as belonging to tribes. However, it is more accurate to characterize them as clan collectives. This is because a clan constitutes a limited group of households united by a common interest in survival, and occupying a collectively owned territory inherited from their putative common ancestors, and whose criterion of membership is not common descent but common geographical origin, or the clan territory called *dabar* in Soqoṭri. Even the Soqotran's definition of clan captures the symbiotic nexus between social organization and environmental context: the term *shaṭr* (meaning: part, section, segment), which is borrowed from Arabic and is attached to a Soqoṭri suffix, gives the following word: *shaṭrihir*, which means strips from a cloth. In effect, it describes the amoeba-like sub-division process of Soqotra's clan formation as well as evokes the image of a cloth being torn apart continuously into ever-thinner strips, partly to accommodate the spatial requirements of grazing herds as part of an extensive form of pastoralism. Indeed, the socio-political organization of the Soqotran community into atomized clans instead of aggregated tribes was structured through the symbiosis between the island's physical geography and the constitution of its social geography: that is, the natural human habitats made available by the island's physical geography have engendered the dispersed nature of its social geography (Figure 2).

Figure 2: Partial view of Hadiboh town.

Regarding the rural–urban population distribution, it is noteworthy that the only two towns on the island have nearly a third (12,407) of the total population of 44,120: Hadiboh (8,545 partly inflated by mainlanders) and Qalansiyah (3,862). The fact that these two towns are located on the northern coast not only confirms that the coastal areas have well over half of the total population but also indicates an inexorable urbanizing, or 'coastalizing' , trend. Moreover, by excluding the population of these two towns, and dividing the remaining total (31,713) with the number of villages (629) included in the census, the average population per village (50.4) corroborates my characterization of Soqotra as a clan-based 'atomistic community' in which the island's landscape is sundered into an archipelago of clans' hamlets. In fact, only about 10% of the non-urban population lives in villages of more than 100 inhabitants. The average population in the overwhelming majority of villages is in double-digit figures, which is partly an effect of the land-extensive herd management system (Figure 3).

Figure 3: Pastoral homestead in the central region.

Finally, the particularities identified for each region are an attempt to provide a succinct, yet tentative, analytical summary in guise of a sociological profile, which expresses the nature of the relationship between residents in different regions and their environment or geographical milieu, as mediated by livelihood practices, which influence population movement, settlement pattern and landscape use. This profile is articulated around the use of two concepts: First, homogeneity, which refers to a population composition that is made up exclusively of related clans engaged in herding activities. This kind of activity is accompanied by a settlement pattern that is centrifugal in orientation because of the extensive (as opposed to intensive) nature of pastoralism in Soqotra. Second, heterogeneity refers to a resident population composed of unrelated clans engaged mostly, but not exclusively, in trade-related activities (e.g. fishing) and are organized in a settlement pattern that is centripetal in orientation because the nature of the economic activities attracts a greater population concentration from diverse locations and clans. These terms are operationalized according to the following criteria: (a) the determination of the homogeneity or heterogeneity of a region's clan composition is based on the ratio of villages to clans: the higher the number of villages *vis-à-vis* the number of clans, the more homogeneous the social composition, or vice versa, (b) whether the pattern of settlements in a region tends toward agglomeration (centripetal) or dispersal (centrifugal), which is determined on the basis of the same criteria for assessing homogeneity and heterogeneity, (c) ranking in population distribution, (d) nature of livelihood activities, and (e) the availability of public amenities: schools, clinics, mosques and government offices. What is clear is that the coastal areas and especially the two urban centres are the migratory poles in an established and accelerating trend of internal migration on the island.

THE PASTORAL FAUNA: A PROVISIONAL INVENTORY

Missing, thus far, from the above discussion is Soqotra's pastoral herds, which is the foundation of the pastoral mode of livelihood that I have been discussing. After more than a decade of international interventions on the island, there are no reliable data on the composition and distribution of livestock around the island. The continuing absence of a reliable census of Soqotra's pastoral herds is a testament to the myopia of both the mainland government and the plethora of international aid agencies, whose focus for more than ten years of intervention and tens of millions of dollars spent on projects was, and remains, almost exclusively on the preservation of endemic species. In fact, the most basic set of information about Soqotra's pastoral system is missing: a) the population and distribution of the livestock herds around the island, b) the extent to which the once effective pastoral management system is still in use or has become dysfunctional, and c) the nature of the ecological effects of the above. For over a generation, the information regarding the livestock population has been based on back-of-the-envelope estimations, as is presented in Table 3 below.

Table 3: Historical data on Soqotra's pastoral herds

Census dates	Animals					Notes and sources
	Goats	Sheep	Cattle	Camels	Donkeys	
1966	19,000	26,000	1,800	350	500	It may have been the first, albeit informal, census, which was carried out by the British colonial officer Brown ([1966]), who candidly confessed that 'field counts are at best only better than guess work'.
Mid-1980s	71,000	17,500	1,900	500	NA	Cited in (Commission of the European Communities EU [2000b]) *Sectoral Report: Agriculture and Livestock*(pp. 2–5) without identifying source.

Mid-1990s	90,000	60,000	12,000	5,000	2,000	These figures are supposed to be from the 1994 agricultural census by the government of Yemen. Given to the author by a staff from Soqotra's branch of the Ministry of Agriculture.
2000	29,300	7,300	2,500	400	NA	Conducted by the European Union Agriculture & Livestock team in 2000 as part of the preparation of the*Soqotra Archipelago Master Plan*. It took place soon after a major drought, which killed an estimated 60% of the livestock.

Elie

Elie Pastoralism 2014 4:16 doi:10.1186/s13570-014-0016-3

These varying figures betray the unreliability of these censuses; nevertheless, they offer a comparative glimpse, even if largely imagined, on the historical trajectory of Soqotra's pastoral herd. While lack of a serious effort on the part of external actors is to be blamed, to be fair, however, this data deficit is exacerbated by the chronic reluctance of Soqotran pastoralists to share the actual numbers of their herds. There are many reasons given for this reluctance: (a) pastoralists really do not know the actual numbers of their herds given their lack of numeracy, (b) the legacy of secrecy about one's assets (whether human or animals) inherited from the Socialist era when the government 'abducted' sons from pastoralist families to be educated in boarding schools away from their villages and parents and (c) *ḩasad wa 'aīn* ('envy and evil eye'), which one pastoralist described in rather hyperbolic term as the 'the Bedouins' creed' (*al-'aqīda al badū*) to refer to a primal fear among pastoralists about talking about their herd size due to the ominous consequences on the fate of their animals. However, the justification for this reluctance that seems to capture the essence of the problem was explained in more modern metaphorical terms by an urbanized former pastoralist as follows: 'It is like asking people how much money they have in the bank'. This is a question that most people would be reluctant to answer. Nevertheless, that reflexive reluctance about revealing herd size could be disarmed in exchange for needed assistance (e.g. veterinary services).

Today, the actual figure for the population of livestock in Soqotra remains elusive, and the most recent attempt to remedy this situation is still unsatisfactory. Nevertheless, the figures obtained from the 2002 unpublished agricultural census are presented in Table 4 for illustrative purpose only. They are cited in the official development plans of the two districts of Soqotra: *An Overview of Development for the District of Hadiboh* (2006:20) and *The Development Report for the District of Qalansiyah* (2007:37). Finally, my use of these figures is due first to the fact that is all there is, and second to their seeming plausibility given that most numbers are not rounded off, which suggests an attempt at actual counting not just estimating, even though the difference with the figures in Table 3 suggests either they are far-fetched or that the previous censuses were partial sampling of the pastoral herds.

Table 4: Census of Soqotra's pastoral herds (2002)

Districts	Animals					Grand total
	Goats	Sheep	Cattle	Camels	Donkeys	
Eastern (Hadiboh)	178,000	26,280c	10,629	66	968	215,943
Western (Qalansiyah)	274,840	75,246	0	0	0	350,086
Total	452,840	101,526	10,629	66	968	566,029

Elie

Elie Pastoralism 2014 4:16 doi:10.1186/s13570-014-0016-3

On the basis of these figures, a few observations can be made:

First is the regional variation in herd composition, and the surprisingly higher number of goats and sheep in Qalansiyah than in Hadiboh district, in spite of the latter's larger territorial expanse and higher human population. However, this is compensated by Hadiboh's more varied herd composition. These figures, whatever their degree of reliability, confirm the cultural distinction between the east and west (i.e. that cattles are neither herded nor eaten in the west), and more importantly, the enforced herd specialization in the west is due to its greater aridity than the east. Also, the larger number of herds in the west suggests a deliberate strategy of claiming territorial grazing rights, if not land ownership, through the presence of specific clan-related herds on the land, in order to make up for the absence of human residents given the west's relatively sparse population. This confirms the fact that pastoralism in Soqotra is first and foremost a mode of territorial occupancy in which the value of the herds is primarily as markers of individual pastoral families as shared 'owners' in the clan collective's territorial commons; and only secondarily a system of livestock production. Furthermore, it might also suggest the deployment of a risk management strategy through the 'maximum-standing-stock' technique of herding where there is a maximization of herd size as an adaptation strategy to the west's chronic deficit of environmental resources. This risk management strategy seeks to ensure the survival of part of the herds in case of resource failure, and not what appears through casual observation as the abandonment of herd management. Second, the livestock population density for the island as a whole is 156 animals per km^2, which is much higher than the human population density of 12.17 persons per km^2. Disaggregated at the level of district, the animal density for each district is as follows: for the Eastern district[c], which occupies less than a third of the island (approx. 995 km^2), the herd density is 351.9 animals per km^2, and for the Western district, which occupies more than two thirds of the island (approx. 2,630 km^2), the herd density is 82.1 animals per km^2 (Figure 4).[d]

[c]In addition, as the number of sheep (7,825) for Eastern district as reported in its development report seemed too low, I have added the figures for sheep (18,455) collected by the 'Community-Based Livestock Development and Marketing Improvement Project', which got reliable figures in exchange for vaccination. However, the project intervened in only a few areas within the Eastern district.

[d]The above estimation is based on the more frequently used figure of the island's

landmass as measuring 3,625km^2. However, the Central Statistical Organization of the Government of Yemen gave the author the following dimensions for each district: East 2,701km^2, and West 1,002.5km^{2}for a total of 3,703.5km^{2}for the island. Also, there are two more figures of the size of Soqotra Island that are in use in the public domain: 3,650 and 3,796km^2.

Figure 4: Cow pastures in the eastern.

Relatedly, the average size of the herd per family/household (with an average of 7.59 persons) in the pastoral areas excluding the urban areas is 295.50. The optimal size of herds has not been determined in the context of Soqotra's grazing resources and the frequency of the island's climatic vagaries (e.g. unpredictable rainfall and drought); therefore, it cannot be ascertained whether this figure exceeds the island's carrying capacity – if such a measure can be meaningfully established. This figure is based on taking the total number of families outside the two towns of the island as listed in the census report in 2004 (Central Statistical Organization CSO [2004]: 137, 141) and dividing them by the total herd figure. Using the same formula, but disaggregated by district, the breakdown of the average number of animals per rural household is as follows: for Hadiboh district, it is 60.56 animals per each of the 3,516 rural families out of a total of 4,658 families, and for Qalansiyah district, it is 530.43 animals per each of the 660 rural families out of a total of 1,123 families. These figures are merely indicative, and their use here is to fill a lamentable informational void due to the collective failure of the mainland government and its international collaborators, as a result of an appreciation deficit of the pastoral fauna as the foundational livelihood activity in Soqotra as well as a key enabler of the island's biodiversity preservation.

CONCLUSION: FROM INCREMENTAL ADAPTATION TO TRANSFORMATIONAL ACCOMMODATION

Given the constantly evolving transition process in Soqotra, there cannot be any ultimate 'conclusion' to this discussion, instead only a provisional assessment can be offered about the nature of changes within the Soqotran community thus far. Indeed, Soqotran pastoralists have been engaged in a series of internal adjustments *vis-à-vis* the externally imposed imperatives that are reconfiguring their communal repertoire of livelihood practices. The socio-cultural transformation of the pastoralists is due to the cumulative effects of the political mobilization strategies and the economic policies of the mainland government since 1967. The result is a series of spatial displacements and social mutations, which are summarized below in terms of the changes in the Soqotran pastoralists' domains of livelihood, residential location and socio-cultural milieu:

The impact of the spatial displacement that took place within Soqotrans' domains of livelihood – that is, the spatial/territorial domains, and the resources contained therein, from which Soqotrans eke out a living – is best characterized as the delinking of livelihood making from its traditional locales. The initial symbiosis between the pastoralist and his grazing grounds, as manifested in the seasonal circularity of the transhumant pattern of *merqīyo* and *mezhīro*, has partially unraveled under the pressure of the island's external entanglements. The end result is a transition, still unraveling, from an environmental dependency linked to the pursuit of human survival through a pastoral mode of subsistence to a cash dependency on non-pastoral activities that are tenuously linked to the cultural continuity of a marginalizing pastoral way of life.

The residential location of Soqotrans has been the most susceptible to external entanglements. The change in residential location was part of an initial process of socializing troglodytic pastoralists, and subsequently of spatial demarcation of the island to accommodate the changing political geography of the different regimes introduced from the mainland. More significantly, the expansion of road construction and public sector post-creation accelerated the shift in the island's population distribution from the hinterland to the coast.

The shift in socio-cultural milieu occasioned by the island's external entanglements is perhaps most significant, since it represents the culmination of a series of state-engendered delocalization initiatives. Indeed, this shift does not merely exemplify a topographical displacement from mountains to coast or from rural to urban, but from one cultural universe to another: from *bādiya*(hinterland) to *madīna* (city). One of the many consequences of this shift is the conversion of the herdsmen into townsmen, as the former seek to accommodate Soqotra's newly imported way of life. The main reason is to ensure the future of the new generation, which entails attending school to learn how to read and write Arabic and English with the ultimate goal of getting a salaried position. For the new generation, the ideal of success is a job with a regular salary that alienates them from their pastoralist upbringing, which has already brought about a profound mutation in Soqotra's pastoral way of life.

In effect, the above discussion of socio-cultural changes in Soqotran pastoralists' lifestyle represents a series of internally driven *incremental adaptations* to a changing economic context. In contrast, there is another source of change that is being generated through an externally enforced*transformational accommodation* of pastoralists as well as non-pastoralist Soqotrans to an island-wide environmental management regime. The primary manifestation of this regime is the 'Zoning Plan for the Soqotra Archipelago: For the Conservation and Sustainable Use of Biodiversity and Natural Resources'. This Zoning Plan was formulated by international consultants and decreed into law by the Yemeni state in 2000. The implementation of this Zoning Plan is managerially accountable to, and financially dependent on, international agencies. It proposes a total encompassing of the island in which not a single square mile of the island is not under some kind of regulatory edict. The latter makes excessive use of the idiom of exclusion, protection and control: lots of sticks but no carrots. Consequently, the plan's virtual reconfiguration of the island's landscape betrays a pervasive disconnect between the demarcated zones and Soqotrans' normal pursuit of their traditional livelihoods. This is tantamount to the symbolic, if not *de facto*, expropriation of their pastoral land as over 75% of the island's surface is under the IUCN strict protection category (national park and nature reserve) in which humans are considered an invasive species! The subsequent inclusion of the island in UNESCO's World Heritage Sites in 2008 enhanced its global recognition. All of these measures have engendered a number of unintended consequences in the form of perverse outcomes, outlined below.

The rise in local expectations from international assistance was subsequently followed by the spread of disappointment with, and chronic doubts about, international donors' projects and promises. The disappointment was due to the implementation of an incongruous nature-preserving, pastoral livelihood-neglecting and modernization-prevention 'sustainable development' model for the island that far exceeds the exigencies of environmental conservation. The formulation of this proscriptive conservation regime was driven more by opportunistic space appropriation by the Zoning Plan designers due to the *carte blanche* given by the national government, rather than an ecological imperative based on a substantiated analysis of the threats to the island's environment. This, in turn, has led to the estrangement of most Soqotrans from international donors' environmental protection initiatives, since these initiatives were chronically unable to demonstrate that biodiversity conservation through the sustainable use of local environmental resources could produce economically viable livelihoods.

The promotion of a tourism economy has quickly degenerated from the ideal of an 'ecologically sensitive low-volume tourism' with pastoralists as its primary beneficiaries to the reality of an ecologically indifferent mass tourism that has benefitted mostly urban-based Soqotrans. Paradoxically, it was the promise of this tourism economy that led the government to construct a 430-km network of asphalted roads that is now seen as a threat to the island's biodiversity. Relatedly, this tourism economy caused a coastal land grab by external actors and exacerbated local disputes over the communal ownership of land. These disputes were animated by local expectations that potential external interests in land as conservation enclaves or tourist business ventures would become a source of rent extraction. This created incentives to monetize communal land holdings through the privatization of property rights.

The spread of a car culture on the island has resulted from the investments of the Soqotran diaspora in the Arabian Gulf in response to income-generating opportunities in the emerging tourism economy. As of January 2014, according to the Director of Traffic Police, there were a total of 1,716 cars on the island, constituting a ratio of one car for every 29 inhabitants. Indeed, the car has become the key INVESTMENT asset of individual participants in the tourism economy as the means of transport for 'ecotourists', environmental researchers and project managers, and thus further contributing to environmental stress. Also, the car has achieved the status of a consumer fetish that is breeding social competition and differentiation within the hinterland. There is now the frequent incongruous site of a relatively new car parked next to an archaic compound, thereby accelerating the demise of the hinterland's communal ethos of 'subsistence egalitarianism'.

In light of the above, conservation practitioners, both local and foreign, should temper their conservationist zeal and recalibrate the Zoning Plan, which has configured the entire island into a 'territorial set-aside' for the sake of the globe's environmental health, while threatening pastoralists with displacement from their livelihood domains. This recalibration process should entail the articulation of a locally adapted sustainable development vision for Soqotra: (a) that prioritizes the role of pastoralists and their herds based on empirically established knowledge about the livestock population, and about the current status of the practice of transhumance, (b) that formulates socio-culturally informed proposals on the economic potentials of pastoralism, and (c) that indigenizes the values guiding the island's environmental protection regime.

ACKNOWLEDGEMENTS

The author acknowledges the useful suggestions of the anonymous reviewers of the *Pastoralism*journal and the appreciation and constructive feedback of the journal editor, Carol Kerven. Also, a special thanks to Dr. Miranda Morris for her critical reading of the manuscript and assistance with the Soqoṭri terms (although I have not always used the diacritical marks suggested). This article is an abbreviated version of a longer chapter in a book manuscript under preparation.

REFERENCES

1. [www.oneworld.org/odi/rfs/r.blench] Blench, RM. 2000. You can't go home again. Extensive pastoral livestock systems: Issues and options for the future. .
2. Brown, GH. 1966. Social and economic conditions and possible development of Socotra. Federal Government Mission Report, Aden.
3. 2004. The final results of the general census for population, housing and establishments 2004. CSO, Sanaa.
4. 2000. Soqotra Archipelago Master Plan. MoPD, Sana'a.
5. 2000. Sectoral Report: Agriculture and Livestock. MoPD, Sana'a.
6. De Sanctis, M, A Adeeb, A Farcomeni, C Patriarca, A Saed, and F Attorre. 2013. Classification and distribution patterns of plant communities on Socotra Island, Yemen. Applied Vegetation Science 16 (1): 148–165.
7. Elie, SD. 2004. Hadiboh: From peripheral village to emerging city. Chroniques Yémenites 12: 53–81.
8. Elie, SD. 2006. The waning of a pastoralist community: An ethnographic exploration of Soqotra as a transitional social formation. D.Phil. dissertation, University of Sussex, UK.
9. Elie, SD. 2009. State-community relations in Yemen: Soqotra's historical formation as a sub-national polity. History and Anthropology 20: 363–393.
10. Elie, SD. 2012. Cultural accommodation to state incorporation in Yemen: Language replacement on Soqotra Island. Journal of Arabian Studies 2: 39–57.
11. Elie, SD. 2012. Fieldwork in Soqotra: The formation of a practitioner's sensibility. Practicing Anthropology 34: 30–34.
12. Elie, SD. 2014. Soqotra's pastoral economy: From core to auxiliary livelihood. Pastoralism: Research, Policy and Practice. in press.
13. Ellen, R. 1982. Environment, subsistence and system: The ecology of small-scale social formations. Cambridge University Press, Cambridge.
14. Ellen, R. 1994. Modes of subsistence: Hunting and gathering to agriculture and pastoralism. In Companion encyclopedia of

anthropology: Humanity, culture and social life, ed. Ingold T, 197–225. Routledge, New York.

15. Khazanov, AM. 1994. Nomads and the outside world. University of Wisconsin Press, Madison.
16. Miller, A, and M Morris. 2004. Ethnoflora of the Soqotra Archipelago. Edinburgh University Press, Edinburgh.
17. Miller, AG, MJ Morris, and RA Atkinson. 2003. Conservation on Soqotra: Ravaging goats or climatic change.
18. Morris, M. 2002. Soqotra Archipelago: Manual of traditional land-use practices. Royal Botanic Garden, Edinburgh.
19. Naumkin, V. 1993. Island of the Phoenix: An ethnographic study of the people of Socotra. : Reading: Ithaca Press.
20. Swift, J, and C Toulmin. 1992. Guidelines and strategies for pastoral development in Africa.

Chapter 7

Heterologous Hybridisation to a Pinus Microarray: Profiling of Gene Expression in Pinus Radiata Saplings Exposed to Ethephon

Joseph G Dubouzet[1], Lloyd Donaldson[1], Michael A Black[2], Les McNoe[2], Vincent Liu[1], and Gareth Lloyd-Jones[1]

[1]Scion, 49 Sala St., Private Bag 3020, Rotorua, New Zealand

[2]Department of Biochemistry, University of Otago, Dunedin, New Zealand

ABSTRACT

Background

The design, construction and application of a *Pinus* microarray platform are described. The oligonucleotide microarray was developed using publicly available *Pinus* cDNA sequences mostly derived from *Pinus taeda* to test whether heterologous hybridisation of microarray probes will generate useful data when hybridised with cRNA constructed from the dominant New Zealand forestry species *Pinus radiata*.

Methods

A comprehensive consensus sequence collection of *Pinus* cDNA sequences was collated into a non-redundant database used for automated design of 60-mer oligonucleotide microarray probes. The microarray slides, manufactured by Agilent Technologies (Palo Alto, California), were used to monitor gene expression in an induction experiment using 2-chloroethylphosphonic acid, common name ethephon and the active ingredient of the plant growth regulator Ethrel (Bayer Crop Science). The transcriptomes from tissues of 2-year old *Pinus radiata* saplings +/− ethephon treatment were compared by hybridisation onto the *Pinus* microarray slides.

Results

Statistically significant differentially expressed genes identified by heterologous hybridisation to the*Pinus* microarray following ethephon induction included the up-regulation of genes in the xylem that were related to the metabolism of phenylpropanoids and flavonoids, and also defence responses, specifically against fungal/insect attack and oxidative stress. Bark, mucilaginous xylem and xylem generated largely mutually exclusive cohorts of genes and Gene Ontology (GO) classes. The results are also interpreted in reference to gross and microscopic morphological changes. Samples of gene responses were validated by quantitative RT-PCR.

Conclusion

These results confirm the successful development of a *Pinus* microarray and demonstrate the utility of the microarray for transcriptomic research in *Pinus radiata* through heterologous hybridisation.

BACKGROUND

Microarray technology has been extensively used to profile gene expression for more than a decade. The ability to perform up to a million parallel hybridisations on a single microarray provides researchers with a very efficient platform for characterisation of genetic processes in target organisms. In addition to characterising the transcriptomic differences generated from an experiment involving one species, the same microarray platform can be used for research involving related species through heterologous hybridisation (Davey et al. [2009]). Microarrays can also be used for studies involving methylation and epigenetics by hybridisation with bisulfite-treated genomic DNA (Reinders et al. [2008]). This versatility in other downstream applications helps to ensure that microarray technology continues to be relevant in crop research, despite the availability and widely recognised superiority of RNA-Seq for transcriptomic analyses.

Pinus radiata D. Don is grown in 1.6 M ha in New Zealand, making it the premier species for silviculture (New Zealand Forest Owners Association [2011]). Microarrays based on comprehensive transcriptomic data can be used to advance pine research, however most examples of *Pinus*microarrays published in the last 10 years are based on very modest collections containing less than 3,500 sequenced genes (Heath et al. [2002]; Yang et al. [2004]; Myburg et al. [2006]; Adomas et al. [2007]; Balk et al. [2007]; Perdiguero et al. [2012]; Li et al. [2012]). A cDNA microarray based on 26,496 sequences for *P. taeda* was recently used to assess the effects of drought in *P. taeda* L. (Lorenz et al. [2011]).

As of July 2013, there were approximately 13,000 publicly available nucleotide sequences from *P. radiata* curated at GenBank (http://www.ncbi.nlm.nih.gov/ *webcite*), but much larger collections of DNA sequences exist from other *Pinus* species, principally *P. taeda,* which had approximately 140,000 sequences. Microarrays can be used for comparative genomic hybridisation among near relatives via heterologous hybridisation (Davey et al. [2009]). Heterologous hybridisation between cDNA probes from *P. taeda,* immobilised on nylon membrane, with cDNA from *P. sylvestris* L. and*Picea abies* (L.) H. Karst has been successful (Van Zyl et al. [2002]). Heterologous hybridisation requires microarray probes that are capable of tolerating a certain degree of mismatch with the target transcripts. Single nucleotide deletions in targets of 60-mer probes can reduce signal intensity by 20-50% whereas single deletions in targets of 20-mer probes essentially eliminated any signal (LeProust [2008]). Probes of 60-mer can still generate useful signals even with 10% mismatched bases between the probe and target gene (Hughes et al. [2001]).

The utility of a microarray is best appreciated by the degree to which the results from this technology confirm or clarify well-characterised biological phenomena. The effect of long-term exposure to an ethylene-generating paste containing 2-chloroethylphosphonic acid (ethephon) on the stem anatomy of *Pinus radiata* saplings was used as a physiological model to test heterologous hybridisation to the *Pinus* microarray. The response of xylogenesis to ethylene has not previously been studied adequately at the molecular level in *Pinus* species. Ethephon breaks down in plant tissues to form ethylene, a major plant hormone with diverse effects on the plant's physiology (Warner and Leopold [1969]). It is known that different *Pinus* species show clear responses to ethephon treatment. Previously, the application of an ethylene-generating paste has been found to induce localised radial stem growth in white pine (*Pinus strobus* L.) (Brown and Leopold [1973]). Ethylene also stimulated growth in bark and xylem tissues of *Pinus halepensis* Mill. (Yamamoto and Kozlowski [1987a]). Ethephon reduced lignification of tracheids in compression wood that developed on the underside of tilted stems of *Pinus densiflora* Siebold & Zucc. (Yamamoto and Kozlowski [1987b]). In *P. halepensis*, ethephon application implicated ethylene in the control of resin duct formation (Yamamoto and Kozlowski [1987a]). Ethephon sprays of 1–2 g L^{-1} effectively reduced height and radial growth of jack pine (*Pinus banksiana* Lamb.) (Schnurr et al. [1996]).

The central assumption of this study is that heterologous microarray hybridisation can be used to generate useful data with probes from closely related species. Microarray probes were designed primarily from *P. taeda* hybridised with cRNA constructed from the close relative *P. radiata*. The creation of a comprehensive consensus sequence database for *Pinus* species, the design and construction of a microarray platform from this database, and the use of this microarray platform to monitor and characterise the transcriptomic response of *P. radiata* saplings to ethephon will be described. The results of heterologous hybridisation were verified by comparing signal data reported from specific probes in the array with physiological observations and, for a limited number of probes, by quantitative reverse transcription polymerase chain reaction (RT-PCR). The lists of differentially expressed genes were analysed using the Database for Annotation, Visualisation and Integrated Discovery (DAVID) Bioinformatics Resources (Huang et al. [2008]) and are discussed in relation to gross and microscopic changes induced by ethephon application.

METHODS

Plant Materials

One-year-old rooted clonal saplings of a single genotype of *Pinus radiata* were purchased from ArborGen Australasia (Whakatane, New Zealand) for this experiment. These were placed in 500 mL pots containing standard potting mix that were then arrayed on an unroofed concrete pavement for another year prior to their use in this research. A total of 60 saplings were required to provide 15 plants per treatment for 4 treatments.

Ethephon Application

Lanolin pastes (0.2 to 0.5 g) containing 0, 1, 5 and 25 mg of ethephon supplied from ETHREL (48% Ethephon, Bayer Crop Science, NZ) were dispensed onto 3 × 6 cm aluminium foil sheets that were wrapped around the basal stem of each experimental sapling, 5–7.5 cm from the soil level. The 1 mg per plant dosage corresponded to the 1% in lanolin dose used in *P. halepensis* (Yamamoto and Kozlowski [1987a]).

RNA Extraction

RNA was extracted from bark, mucilaginous xylem (containing the cambium) and xylem samples. Bark refers to the outermost layer of tissues of the stem beyond the mucilaginous developing xylem. Xylem refers to the woody fibrous tissue left after scraping off the mucilaginous developing xylem. RNA was extracted from the 0 and 1 mg treatments for transcriptomic analysis. The 5 and 25 mg ethephon treatments at 8 weeks after treatment (WAT) elicited the greatest physiological responses but transcriptomic analyses were not attempted for these treatments due to inconsistent RNA extraction possibly due to the adverse effect of high resin content. Three replicates consisting of 5 saplings per replicate were destructively sampled 1 week after ethephon treatment and an identical batch was sampled 8 weeks after treatment. Thus each tissue was sampled from each of 15 saplings for a given treatment. After wiping off the residual lanolin from the bark surface with tissue paper, the bark was cut and peeled off, the mucilaginous xylem was scraped off with the blunt edge of a box cutter, and the remaining woody xylem was cut off from the rest of the stem. The respective tissues were wrapped in water-proof paper, flash-frozen in liquid nitrogen and stored at −80°C prior to RNA extraction. Total RNA from xylem was extracted using Concert reagent (Invitrogen; Carlsbad, CA) whilst RNA extracted from bark was obtained using Concert reagent followed by purification with the Plant RNA Isolation kit (Agilent). RNA from the mucilaginous xylem was extracted using the Plant RNA Isolation kit.

Assembly and Annotation of the Heterologous *Pinus* Consensus Transcript Database

In mid-2010, there were 18,900 Unigene entries for *P. taeda* in the US National Center for Biotechnology Information (NCBI) database (ftp://ftp.ncbi.nih.gov/repository/UniGene/Pinus_taeda/Pta.seq.uniq.gz). The Plant Genome Database (http://www.plantgdb.org/) had 77,500 Putative Unique Transcripts (PUT), (http://www.plantgdb.org/download/download.php?dir=/Sequence/ESTcontig/Pinus_taeda/current_version/Pinus_taeda.mRNA.PUT.fasta.bz2). Harvard's Computational Biology and Functional Genomics Laboratory (http://compbio.dfci.harvard.edu/cgi-bin/tgi/gimain.pl?gudb=pine) had 39,800 tentative consensus (TC) sequences (ftp://occams.dfci.harvard.edu/pub/bio/tgi/data/Pinus/PGI.release_8.zip). These sequences were combined into a FASTA file generated using BioEdit software (Hall [1999]) and then reassembled using the default parameters for sequence assembly using CAP3 software (http://seq.cs.iastate.edu/CAP3/cap3.linux.i686_xeon64.tar) (Huang and Madan [1999]). The final database consisted of 68,225 sequences, with 39,869 assembled contigs in common between at least two of the three databases and 28,356 'singletons' unique to one database. The assembled sequences ranged from 50 to 6175 base pairs (bp), with an average length of 811 bp. The distribution of the assembled sequences is shown in Figure 1. The reassembled database, henceforth referred to as the All-*Pinus* database, was used for the design of 60-mer probes for the *Pinus* microarray.

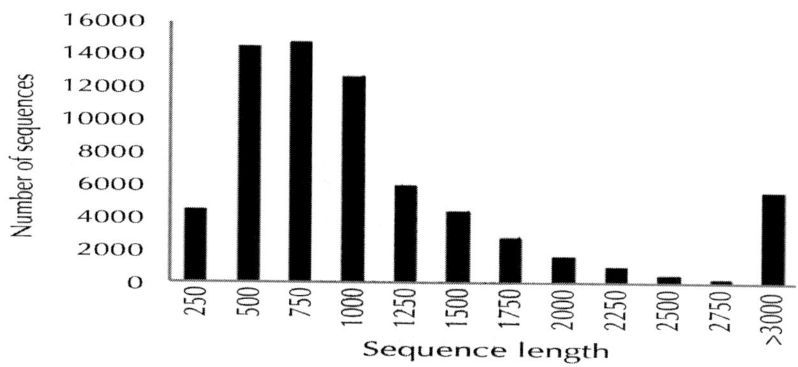

Figure 1: Sequence size distribution in the All-*Pinus*database.

The most similar protein for each sequence assembly was determined using the BLASTX algorithm as implemented with the BioEdit software, using the default criteria and a maximum E-value set at 0.001. The closest protein was identified as the one with the most significant E-value among those identified by BLASTX. The actual average E-value for the whole database was 1.9×10^{-5}. With these criteria, 40,238 (59%) sequences were assigned a putative protein ID, of which 14,838 were unique. A cut-off E-value set at 1×10^{-5} was used in the annotation of Conifer DBMagic (Lorenz et al. [2012]) and a smaller number of putative protein IDs (36,995 sequences (54%)) within the All-*Pinus* database sequences were identified when this cut-off E-value was used. The annotations for these protein IDs were downloaded from DAVID v6.7, (http://david.abcc.ncifcrf.gov/) (Huang et al. [2009]). The top 40 sequences identified by comparison with the Swiss-Prot database of annotated protein sequences and the numbers of transcript variants that code for essentially the same protein are shown in Table 1.

Table 1: Top 40 most common proteins identified in the All-*Pinus*database

SwissProt ID	Description and original species	Number of sequences	Proportion
Q75JC9	Uncharacterised protein DDB_G0271670 OS=Dictyostelium discoideum	33	0.20%
O94317	Uncharacterised serine-rich protein C215.13 OS=Schisosaccharomyces pombe	27	0.16%
Q17RH7	Putative protein TPRXL OS=Homo sapiens	26	0.16%
P26986	2S seed storage-like protein OS=Picea glauca	19	0.12%
Q8LGG8	Universal stress protein A-like protein OS=Arabidopsis thaliana	18	0.11%
P37219	Abscisic stress-ripening protein 2 OS=Solanum lycopersicum	15	0.09%
P19037	18.1 kDa class I heat shock protein OS=Arabidopsis thaliana	14	0.09%
O65351	Subtilisin-like protease OS=Arabidopsis thaliana	13	0.08%
Q6J163	Auxin-induced protein 5NG4 OS=Pinus taeda PE=2 SV=1	13	0.08%
P13240	Disease resistance reprotein 206 OS=Pisum sativum	12	0.07%
Q00451	36.4 kDa proline-rich protein OS=Solanum lycopersicum	12	0.07%
Q40392	TMV resistance protein N OS=Nicotiana glutinosa	12	0.07%
Q50EK4	Cytochrome P450 750A1 OS=Pinus taeda	11	0.07%

Q8VZ13	Uncharacterised protein At1g08160 OS=Arabidopsis thaliana	11	0.07%
Q9S7C9	Putative DNA-binding protein ESCAROLA OS=Arabidopsis thaliana	11	0.07%
C0LGT6	LRR receptor-like serine/threonine-protein kinase EFR OS=Arabidopsis thaliana	10	0.06%
O80337	Ethylene-retranscription factor 1A OS=Arabidopsis thaliana	9	0.06%
P27061	Acid pho1 OS=Solanum lycopersicum	9	0.06%
Q766C3	Aproteinase nepenthesin-1 OS=Nepenthes gracilis	9	0.06%
Q8BTI8	Serine/arginine repetitive matrix protein 2 OS=Mus musculus	9	0.06%
O23530	Protein SUPPRESSOR OF npr1-1, CONSTITUTIVE 1 OS=Arabidopsis thaliana	8	0.05%
P17407	21 kDa protein OS=Daucus carota	8	0.05%
P50160	Sex determination protein tasselseed-2 OS=Zea mays	8	0.05%
Q02817	Mucin-2 OS=Homo sapiens	8	0.05%
Q39224	Protein SRG1 OS=Arabidopsis thaliana	8	0.05%
Q41001	Blue copper protein OS=Pisum sativum	8	0.05%
Q50EK3	Cytochrome P450 704C1 OS=Pinus taeda	8	0.05%
Q55909	TVP38/TMEM64 family membrane protein slr0305 OS=Synechocystis	8	0.05%
Q5ZJC8	mTERF domain-containing protein 1, mitochondrial OS=Gallus gallus	8	0.05%

Q766C2	Aproteinase nepenthesin-2 OS=Nepenthes gracilis	8	0.05%
Q84UV8	Bifunctional monodehydroascorbate reductase and carbonic anhydrase nectarin-3 OS=Nicotiana langsdorffii x Nicotiana sanderae	8	0.05%
Q95JC9	Basic proline-rich protein OS=Sus scrofa	8	0.05%
Q9LHE3	Protein APROTEASE IN GUARD CELL 2 OS=Arabidopsis thaliana	8	0.05%
Q9SB61	ZF-HD homeobox protein At4g24660 OS=Arabidopsis thaliana	8	0.05%
Q9SEV0	Anthocyanidin reductase OS=Arabidopsis thaliana	8	0.05%
Q9SJL0	UDP-glycosyltransferase 86A1 OS=Arabidopsis thaliana	8	0.05%
Q9SZN7	Heavy metal-associated isoprenylated plant protein 26 OS=Arabidopsis thaliana	8	0.05%
Q9UQ35	Serine/arginine repetitive matrix protein 2 OS=Homo sapiens	8	0.05%
Q9ZNU2	NAC domain-containing protein 18 OS=Arabidopsis thaliana	8	0.05%
P05790	Fibroin heavy chain OS=Bombyx mori	7	0.04%

Dubouzet *et al.*

Dubouzet *et al. New Zealand Journal of Forestry Science* 2014 44:21 doi:10.1186/s40490-014-0021-8

Pre-Testing the All-*Pinus* Sequence Database

The 7,538 expressed sequence tags (ESTs) from *P. radiata* found in the US National Center for Biotechnology Information (NCBI) database as of 2010 were used to query the 68,225 sequences in the All-*Pinus* database via BLASTN algorithm (Altschul et al. [1997]). The BLASTN algorithm was set at a cut-off value of $1.0 \times E^{-10}$. The alignments that generated less than the highest score for each EST were eliminated. The average data from BLASTN (630 bp aligned length, 99.2% identity, 619 bp actual length, 12.7 unaligned bases, 3.7 mismatched bases and 0.8 gaps) reinforced the potential of heterologous hybridisation of *P. radiata* sequences to a *Pinus* microarray developed from the All-*Pinus* database.

Microarray Probe Design and *in Silico* Validation

Candidate probes were selected based on base-composition profiles optimised through eArray (https://earray.chem.agilent.com/earray/). Three 60-mer sense probes were selected across the length of the template sequence. The 191,500 candidate probes were trimmed to 175,614 to fit the 180 k format required by the Agilent software. Probes were eliminated based on high similarity and hybridisation quality as predicted by eArray. The remaining probes were used to query the original All-*Pinus* database by BLASTN, which was set at a cut-off value of $1.0 \times E^{-10}$. The results indicated that 123,311 probes (70.2%) for 49,081 genes (71.9%) would give a target-specific signal. The remaining probes were expected to hybridise to sequences such as paralogues and splice variants that have different accession numbers but are essentially different versions of the same protein. The probe sequences were used to query the 7538 ESTs from *P. radiata* in the GenBank sequence database provided by the NCBI. The average identity, aligned length, non-identical bases and gaps were 99.6%, 59.9 bp, 0.2 bp and 0.0 bp respectively. These results affirmed the potential utility of the All-*Pinus* database for generating heterologous *Pinus* microarray probes for analysing transcriptomes derived from *P. radiata*.

Microarray Analysis

Paired hybridisation for each microarray comprised treatment versus control of each tissue from the same time point, for example; 'bark, 0 mg ethephon, 1 WAT' versus 'bark, 1 mg ethephon, 1 WAT', and so forth. Total RNA samples of 50–100 ng were used for labelling with the fluorescent cyanine dyes Cy3 or Cy5 using the Agilent Low input Quick Amp labelling kit. After labelling, unincorporated dye was removed using RNeasy (Qiagen, Valencia, CA) mini-spin columns. Purified samples were assessed for yield and specific activity using a Nanodrop ND-1000 UV–VIS spectrophotometer (Thermo Fisher Scientific, Wilmington, DE). Samples that gave yields of >0.8 μg cRNA and specific activity (Concentration of Cy3/Concentration of cRNA ×1000) >10 were used for arrays. Labelled samples were hybridised to arrays at 65°C for 17 hours at 10 rpm in an Agilent G2545A hybridisation oven using 2× GEx Agilent hybridisation buffer. Slides were washed using Agilent wash buffers. Array scanning and feature extraction was performed using the Agilent G2565CA microarray scanner as outlined in the Agilent protocol for Two-colour Microarray-Based Gene Expression Analysis, Version 6.5, May 2010.

Data were imported into R software using the limma package (Smyth [2005]), transformed to the log scale (base 2), and normalized via loess smoothing with no background correction. Statistical analysis of data was performed using the linear models approach implemented in limma, using the M values for each array as the dependent data. The empirical Bayes shrinkage-based approach to per-probe variance estimation was used to generate a moderated t-statistic for each probe (Smyth[2005]). This statistic was then used to determine the statistical significance of the differences for each gene at each time/tissue combination. Probability (*P*) values were calculated for each comparison, and used to define the False Discovery Rate threshold by the Benjamin-Hochberg method (Benjamini and Hochberg [1995]).

Probes with adjusted *P*-values greater than 0.01, log2 fold changes between −1 and 1, and probes with no Uniprot homologs were eliminated. The remaining significantly expressed probes, divided into up- and down-regulated subgroups, were uploaded onto the DAVID portal (http://david.abcc.ncifcrf.gov/home.jsp) for singular enrichment analysis in comparison with the list of protein sequences in the All-

Pinus database. The number of protein IDs associated with each gene ontology classification that were significantly different at $p<0.01$ and displayed a conservative False Discovery Rate less than 10% are shown in Table 2.

Table 2: Significantly enriched gene ontologies associated with biological processes as deduced by DAVID from a list of differentially expressed genes identified by limma analysis

Category	Term	Count	Pvalue	Fold enrichment	BH	FDR
Mxylem 1						
GO:0010286	Heat acclimation	3	0.00	34.03	0.48	4.39
GO:0006952	Defence response	16	0.00	5.39	0.00	0.00
Xylem 1						
GO:0009813	Flavonoid biosynthetic process	15	0.00	18.17	0.00	0.00
GO:0009812	Flavonoid metabolic process	15	0.00	17.54	0.00	0.00
GO:0009699	Phenylpropanoid biosynthetic process	18	0.00	12.99	0.00	0.00
GO:0009698	Phenylpropanoid metabolic process	18	0.00	9.32	0.00	0.00
GO:0042398	Cellular amino acid derivative biosynthetic process	20	0.00	8.37	0.00	0.00
GO:0019438	Aromatic compound biosynthetic process	19	0.00	7.08	0.00	0.00
GO:0006575	Cellular amino acid derivative metabolic process	20	0.00	5.06	0.00	0.00
GO:0042542	Response to hydrogen peroxide	7	0.00	5.05	0.10	3.57
GO:0019748	Secondary metabolic process	21	0.00	4.88	0.00	0.00
GO:0006952	Defence response	15	0.00	3.36	0.01	0.20
Bark 8						
GO:0009853	Photorespiration	5	0.00	17.49	0.01	0.23
GO:0019253	Reductive pentose-phosphate cycle	4	0.00	15.64	0.05	2.73

GO:0015977	Carbon utilisation by fixation of carbon dioxide	4	0.00	13.99	0.07	3.78
GO:0019685	Photosynthesis, dark reaction	4	0.00	13.29	0.07	4.39
GO:0042744	Hydrogen peroxide catabolic process	9	0.00	8.31	0.00	0.02
GO:0009813	Flavonoid biosynthetic process	7	0.00	8.31	0.01	0.26
GO:0043094	Cellular metabolic compound salvage	5	0.00	8.11	0.07	4.47
GO:0070301	Cellular response to hydrogen peroxide	9	0.00	8.08	0.00	0.02
GO:0009812	Flavonoid metabolic process	7	0.00	8.02	0.01	0.31
GO:0042743	Hydrogen peroxide metabolic process	9	0.00	7.98	0.00	0.02
Mxylem 8						
GO:0009813	Flavonoid biosynthetic process	34	0.00	7.53	0.00	0.00
GO:0009812	Flavonoid metabolic process	35	0.00	7.48	0.00	0.00
GO:0009699	Phenylpropanoid biosynthetic process	40	0.00	5.27	0.00	0.00
GO:0009698	Phenylpropanoid metabolic process	45	0.00	4.26	0.00	0.00
GO:0009626	Plant-type hypersensitive response	8	0.00	4.13	0.10	3.62
GO:0034050	Host programmed cell death induced by symbiont	8	0.00	4.13	0.10	3.62
GO:0009739	Response to gibberellin stimulus	11	0.00	3.50	0.05	1.34
GO:0042398	Cellular amino acid derivative biosynthetic process	41	0.00	3.14	0.00	0.00
GO:0019438	Aromatic compound biosynthetic process	44	0.00	3.00	0.00	0.00
GO:0009415	Response to water	18	0.00	2.94	0.01	0.15
Xylem 8						
GO:0001906	Cell killing	6	0.00	9.31	0.01	0.38

GO:0031640	Killing of cells of another organism	6	0.00	9.31	0.01	0.38
GO:0009812	Flavonoid metabolic process	29	0.00	8.54	0.00	0.00
GO:0009813	Flavonoid biosynthetic process	28	0.00	8.54	0.00	0.00
GO:0009225	Nucleotide-sugar metabolic process	7	0.00	7.03	0.01	0.45
GO:0006032	Chitin catabolic process	9	0.00	6.68	0.00	0.05
GO:0016998	Cell wall macromolecule catabolic process	9	0.00	6.15	0.00	0.09
GO:0006026	Aminoglycan catabolic process	9	0.00	6.15	0.00	0.09
GO:0009699	Phenylpropanoid biosynthetic process	32	0.00	5.81	0.00	0.00
GO:0009698	Phenylpropanoid metabolic process	42	0.00	5.47	0.00	0.00

Mxylem, mucilaginous xylem; 1 and 8 refer to the number of weeks after application of Ethephon in lanolin; BH, Benjamini-Hochberg correction; FDR is false discovery rate.

Dubouzet *et al.*

Dubouzet *et al.* *New Zealand Journal of Forestry Science* 2014 44:21 doi:10.1186/s40490-014-0021-8

Quantitative RT-PCR Verification of Microarray Signals

Ten probes were selected that generated stable signals greater than background across all treatment combinations after heterologous hybridisation of the *Pinus* microarray; the probes included Contig12730c, which showed strong stable signals that did not change with ethephon/sampling time for use as a control. Forward and reverse primers were designed to amplify >45 bp fragments within the 60-mer probe using OligoAnalyzer 3.1 (http://www.idtdna.com/analyzer/Applications/OligoAnalyzer/). Primer details are shown in Table 3 for the successful primer pairs. Aliquots from the original samples that contained 200 to 1300 ng RNA were reverse transcribed using qScript (Quanta Biosciences, MD, USA) in 20 µl reactions. A 1:20 dilution was added as template for a qPCR mix using Lightcycler® Faststart DNA Masterplus SYBR Green i (Roche, USA). A Lightcycler® Carousel-based system (Roche, USA) was used for thermal cycling. A 10 min denaturation step was followed by 45 thermal cycles consisting of a 10 sec denaturation step at 95°C, annealing step of 4 sec at 60°C and an extension step of 3 sec at 72°C. This was followed by melting point analysis from 65°C to 95°C at a ramp rate of 0.2°C sec^{-1}. Fold-change calculation, using the expression level of Contig12730c as the constant divisor, was performed with a formula (3.5) for relative quantification (Pfaffl [2004]).

Table 3: Microarray probe targets for quantitative RT-PCR validation

Key process	Gene	Probe ID	5 Primer	3′-Primer	PCR validation
Catechin Biosynthesis	Anthocyanidin reductase	Contig38855a	TGGCATCGACGAAATCTACG	TTTCGCGGATTCTACAGCGT	Yes
Isoflavonoid biosynthesis	Isoflavone reductase	Contig32761c	CCCGACGTTGAATACACATCAGT	ACGGCTCAAATAACGCTCGACT	Yes
Mono terpenoids	Pinene synthase	Contig819c	CTAACTTAGGTCTCTGATTT	GATATCTTACCCACAACATAGC	Yes
Unknown	Unknown	Contig24580b	AAATATAATCTCTGTCATCTGCG	AAGCGAATGGTCATGGTC	Yes
Cell Cycle	Cyclophilin	Pta121982552a	AGAGTCGATCTACGGCGAGA	CTTTACAAAGTTCTCATCGGCA	Yes

Dubouzet *et al.*

Dubouzet *et al. New Zealand Journal of Forestry Science* 2014 44:21 doi:10.1186/s40490-014-0021-8

Microscopy

The gross physiological effects of ethephon were analysed by microscopic analysis of sectioned stems of all treatments. Stem cross-sections were prepared with a sledge microtome and images of the cut ends of stems were obtained using a stereo microscope. Stems were fixed in FAA (5% formalin, 5% glacial acetic acid, 90% ethanol v/v) fixative for 1 month and then examined via confocal microscopy and autofluorescence. Images were obtained with simultaneous excitation at 488 and 561 nm.

RESULTS

Transcriptomic Analysis of Ethephon Induction in *Pinus Radiata* Saplings

Pinus radiata transcriptomes were analysed by heterologous hybridisation to the *Pinus* microarray with cRNA generated from different tissues extracted from ethephon-treated and control plants. The transcriptomes in bark, mucilaginous xylem and xylem 1 and 8 WAT with 0.2 g of lanolin containing 0 or 1 mg ethephon were compared by two-colour hybridisation. Limma analysis of the data revealed that the mucilaginous xylem had 66 and 23,084 significantly different signals (relative to the pertinent control) 1 and 8 WAT with ethephon, respectively; similarly, bark had 127 and 1761 (1 and 8 WAT) whilst xylem had 1,352 and 12,718 probes exhibiting significantly different signals (1 and 8 WAT).

The relationships between the differentially expressed probes among tissues and across time are shown in the Venn diagrams in Figure 2. At 1 WAT, the mucilaginous xylem and the bark displayed comparable numbers of up- and down-regulated genes, but only one of these was co-expressed. More genes in the xylem showed significant down-regulation due to ethephon but only a very small fraction were co-expressed in the mucilaginous xylem (12 or 3.9%) or bark (5 or 1.6%). Despite the very low numbers of differentially expressed genes in the bark at 1 WAT, at least 50% of these were still expressed in the bark at 8 WAT. A lower proportion (25-34%) of genes that were differentially expressed in the xylem at 1 WAT was also found seven weeks later.

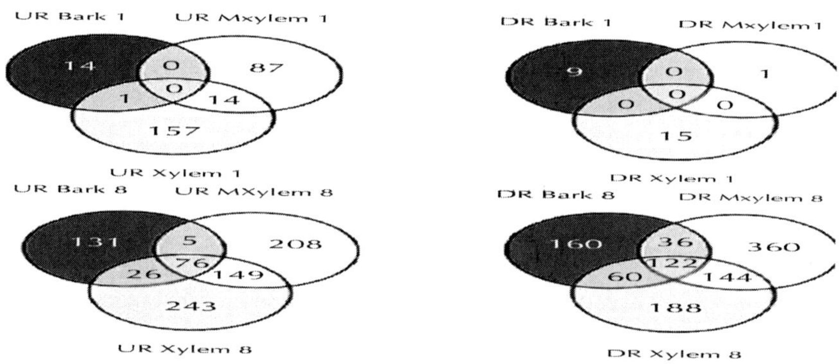

Figure 2: Number of differentially expressed gene cohorts in the various sample tissues. The value in the overlapping portions of the circles corresponds to the number of co-expressed gene cohorts in each treatment pair. Legend: UR, upregulated; DR, downregulated; Mxylem, mucilaginous xylem; 1 and 8 weeks after treatment.

Differentially expressed (fold change >2, adj-$p<0.05$) genes were systematically mapped to the associated biological annotation (e.g., Gene Ontology Terms, KEGG pathways) using DAVID, and the most over-represented (enriched) biological annotation were statistically highlighted. Enrichment analysis increases the likelihood of identifying the biological processes most relevant to the experiment (Huang et al. [2008]). The Functional Annotation Table of Gene Ontologies with False Discovery Rates less than 0.1, generated through the DAVID portal, are shown in Table 2. Ethephon had no significant effect on the bark transcriptome one week after application. The xylem transcriptome had a greater response to ethephon than the bark transcriptome resulting in an increase in the activity of genes associated with cellular amino acid derivatives and phenylpropanoid biosynthesis in xylem at 1 WAT and 8 WAT. Ethephon initially depressed defence responses in the xylem but, after eight weeks, ethephon application activated a wide range of genes related to plant defence. These included genes involved in phenylpropanoid and flavonoid biosynthesis, chitin detection and degradation, oxidation-reduction, and toxin catabolism. Lignin metabolic processes were significantly down-regulated by ethephon treatment.

In the bark, genes associated with structural constituents of the cell wall (GO:0005199) were activated by ethephon. The down-regulation of light-responsive genes in the bark was expected, but the suppression of genes associated with abiotic stress was unexpected.

Validation of Microarray Signals by Quantitative RT-PCR

To test the validity of signal quantitation by the *Pinus* microarray, 60-mer probes were selected and corresponding transcripts in cDNA transcribed from the original total RNA extracts used for microarray hybridisation were quantified by RT-PCR. Five primer pairs designed to amplify a 45–60 bp segment within each probe (Table 4) generated fold change estimates that were in general agreement with the direction (up- or down-regulation) of the raw signal data from the microarrays (Figure 3). The sixth probe used as a constitutive background control (contig 12730c) gave a consistent signal. Four primer pairs failed to amplify a corresponding target which is a consequence of the higher specificity required for PCR amplification which will not always be compatible for the detection of heterologous targets.

Table 4: Significantly suppressed gene ontologies associated with biological processes as deduced by DAVID from a list of differentially expressed genes identified by limma analysis

Category	Term	Count	Pvalue	Fold enrichment	Benjamini	FDR
Bark 8						
GO:0048444	Floral organ morphogenesis	4	0.00	15.78	0.07	2.50
GO:0030244	Cellulose biosynthetic process	7	0.00	9.56	0.01	0.10
GO:0015833	Peptide transport	6	0.00	8.52	0.04	0.88
GO:0006857	Oligopeptide transport	5	0.00	8.07	0.10	4.56
GO:0030243	Cellulose metabolic process	8	0.00	5.07	0.05	1.41

GO:0009699	Phenylpropanoid biosynthetic process	13	0.00	4.91	0.00	0.02
GO:0009250	Glucan biosynthetic process	8	0.00	4.58	0.07	2.58
GO:0006073	Cellular glucan metabolic process	13	0.00	4.35	0.00	0.06
GO:0009698	Phenylpropanoid metabolic process	16	0.00	4.34	0.00	0.01
GO:0009414	Response to water deprivation	8	0.00	4.12	0.10	4.75
Mxylem 8						
GO:0046274	Lignin catabolic process	17	0.00	6.25	0.00	0.00
GO:0046271	Phenylpropanoid catabolic process	17	0.00	6.25	0.00	0.00
GO:0048444	Floral organ morphogenesis	6	0.00	5.15	0.14	4.91
GO:0030244	Cellulose biosynthetic process	16	0.00	4.75	0.00	0.00
GO:0009808	Lignin metabolic process	31	0.00	4.35	0.00	0.00
GO:0051258	Protein polymerisation	25	0.00	4.29	0.00	0.00
GO:0042219	Cellular amino acid derivative catabolic process	19	0.00	3.96	0.00	0.00
GO:0007018	Microtubule-based movement	28	0.00	3.54	0.00	0.00
GO:0009698	Phenylpropanoid metabolic process	56	0.00	3.30	0.00	0.00
GO:0009809	Lignin biosynthetic process	13	0.00	3.14	0.03	0.68
Xylem 8						
GO:0046274	Lignin catabolic process	8	0.00	4.40	0.11	2.32
GO:0046271	Phenylpropanoid catabolic process	8	0.00	4.40	0.11	2.32
GO:0006857	Oligopeptide transport	8	0.00	4.20	0.14	3.14
GO:0009808	Lignin metabolic process	19	0.00	3.99	0.00	0.00

GO:0009698	Phenylpropanoid metabolic process	41	0.00	3.61	0.00	0.00
GO:0009809	Lignin biosynthetic process	10	0.00	3.61	0.11	1.96
GO:0009699	Phenylpropanoid biosynthetic process	29	0.00	3.56	0.00	0.00
GO:0009813	Flavonoid biosynthetic process	16	0.00	3.30	0.01	0.09
GO:0016137	Glycoside metabolic process	10	0.00	3.21	0.19	4.74
GO:0009812	Flavonoid metabolic process	16	0.00	3.18	0.01	0.15

Legend as in Table 2.

Dubouzet *et al.*

Dubouzet *et al. New Zealand Journal of Forestry Science* 2014 44:21 doi:10.1186/s40490-014-0021-8

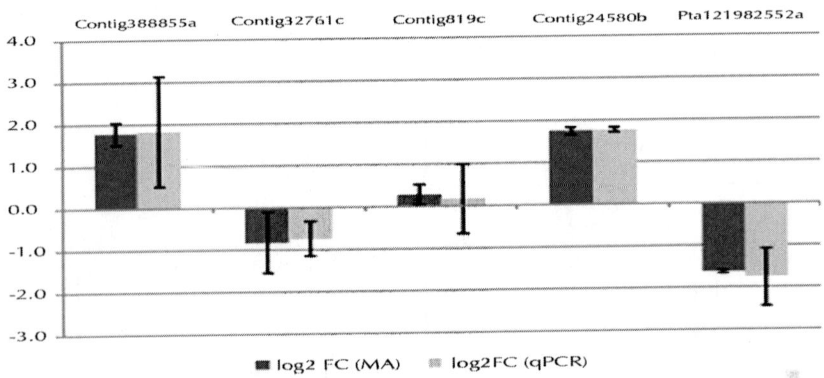

Figure 3: Quantitative RT-PCR verification of*Pinus*microarray signals. Log fold-change values are indicated in the Y axis; negative values correspond to down-regulated transcripts. Bars correspond to standard deviation. Legend: LogFC, Log Fold Change; MA, microarray.

Anatomical Changes after Long-Term Ethephon Exposure

Most of the visible response to ethephon was limited to the treated area on the bark. There was little effect on stem girth and appearance 1 cm above or below the lanolin band (Figure 4). After 8 weeks of treatment, the 1 mg dose induced an increase in girth (16%) relative to the girth immediately above and below the treated area. Increased stem diameter was evident in the ethephon-treated area but not in the xylem. Most of the increased growth was manifested in increased amounts of phloem and cortex in the bark that displayed deep fissures and cracks, and sap bleeding that turned the normally colourless lanolin white. Higher ethephon dosages (5 mg and 25 mg treatments) induced more resinification and generated greater levels of opacity and whitening of the lanolin due to reaction with bark exudates, bark splitting and necrosis and reduced stem growth (data not shown). The ethephon-treated stems exhibited an abnormal layer of unlignified tissue in the outer xylem, which is indicated by an asterisk in Figure 4B. The number and size of the resin canals were unaffected by ethephon treatment. After 1 month of fixation in FAA, unstained sections examined for autofluorescence indicated the presence of strongly red fluorescent extractives in both wood and bark from ethephon-treated plants (Figures 4C and 5). The red fluorescent parenchymatous cells, scattered throughout the outer xylem, are poorly lignified (Figure 5).

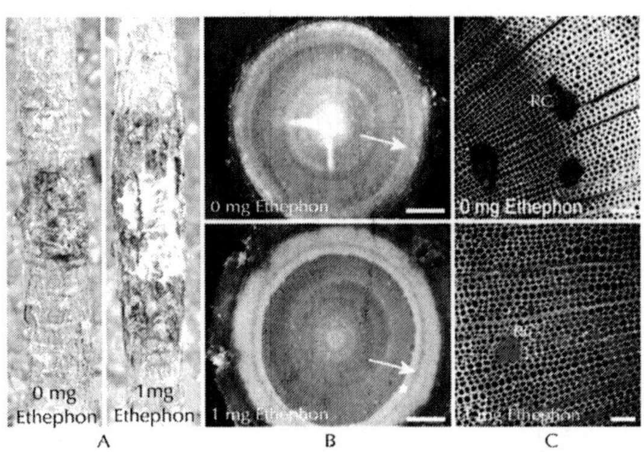

Figure 4: Sapling morphology 8 weeks after application of ethephon. A. Resin leakage from cracks in the bark of stems treated with ethephon, causing some

of the lanolin to turn white. B. increased girth and cracked bark induced by ethephon. Arrows indicate the mucilaginous xylem in each stem while the asterisk indicates an unlignified layer of parenchymatous tissue in the outer xylem in the 1 mg ethephon treatment. Note no appreciable difference in diameter of the lignified xylem or in the number and size of resin ducts (round light-brown spots) was observed between different treatments. Scale bars = 2 mm. C. Comparison of resin canals (RC) after FAA fixation of xylem exposed to 0 mg or 1 mg ethephon. Resin canals and tracheids in xylem of ethephon-treated stem are suffused with unidentified extractives that show red fluorescence on excitation at 488 and 561 nm. Scale bars = 100 μm.

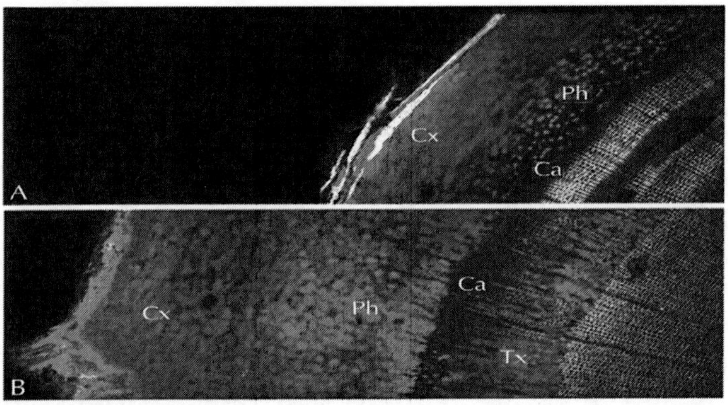

Figure 5: Comparison of outer stem anatomy between control (A) and 5 mg ethephon treatment (B). Both Phloem (Ph) and Traumatic xylem (Tx) appear abnormal and contain increased extractives. The extractives are concentrated in poorly lignified parenchymatous cells in the xylem. Cortex (Cx) and Cambium (Ca) are as indicated. The images are at the same scale. Scale bar = 1 mm.

DISCUSSION

Microarray Construction and Validation

Next Generation Sequencing (NGS) techniques can be used to produce gene libraries for the construction of microarrays (Hiremath et al. [2011]), however generating numerous fully sequenced and annotated transcriptomes is expensive and time consuming. It is more efficient to capture a comprehensive picture of gene activity of an organism through the combining and normalisation of existing transcriptomes derived from various tissues sampled temporally and from a wide variety of natural and experimental conditions. Extensive sequence databases exist for *Pinus taeda* that were generated from diverse experimental conditions by many research groups; less extensive databases are also available for other *Pinus* species. Pre-existing public gene databases for *Pinus* species were used to construct the *Pinus* microarray described in this paper. The *Pinus* data were processed using CAP3 for reassembly of contigs, to remove redundancy and to generate consensus contigs (Huang and Madan [1999]; Min et al. [2009]; Feldmesser et al. [2013]; Huang et al. [2012]) for the All-*Pinus* database. *In silico* validation by BLASTN analysis of the All-*Pinus* database and the probes constructed from it revealed >99% identity of aligned sequences with reference to a set of publicly available *P. radiata* sequences. These *in silico* results were corroborated by the results obtained after *P. radiata* cRNA hybridisation.

General Observations on the Transcriptomes Induced by Ethephon

Large numbers of genes were identified with highly significant levels (adjusted *P*-value < 0.01) of differential expression induced by ethephon treatment, and these were dependent on both tissue type and duration of ethephon exposure (Figure 2). Ethephon is a chemical that diffuses into the underlying plant tissues following application then generates the potent plant hormone ethylene in slightly acid or basic conditions in the cytoplasm (Lieberman [1979]) where the effect of ethephon on endogenous ethylene biosynthesis is transient. This effect occurred in less than 6 days in the case of apple cuttings (Sun and Bassuk [1993]) although the effect of ethephon on *P. radiata* after 1 week was minimal. Once formed, however, ethylene triggered gene expression from each type of tissue at 8 WAT, and these were different in composition (Figure 2) and function (Table 2). Some (25-50%) of the genes expressed at 1 WAT were found to be differentially expressed at 8 WAT, indicating that these genes may form the core response of these tissues to ethephon. Amongst these were at least 8 gene ontology classes that showed bidirectional responses (both up- and down-regulated) to ethephon in the xylem (Table 2). Overall, ethephon acted as a suppressor rather than as a promoter of gene transcription.

In *P. radiata* saplings, the mucilaginous xylem exists as a thin layer located between bark and xylem and includes the undifferentiated 1–2 cell thick cambium layer, expanding cells and a few early secondary walls classifiable as xylem but not phloem (Cato et al. [2006]). This mucilaginous layer was practically absent in ethephon-treated stems by 8 WAT such that the recovered tissues contained proportionally more fibrous material comprising young xylem. At 8 WAT, extensive contamination of the 'tissue scrapes' with more mature xylem tissues from ethephon-treated plants is the most probable reason for the unusually high number (>23,000) of probes showing significantly different signals when compared with corresponding control plants from which 'tissue scrapes' uncontaminated by xylem were obtained. For this reason, the data from this treatment will not be discussed further, but serve as a reminder of the need for robust tissue typing and tissue collection if between-tissue comparisons are to be attempted. When comparing different studies it is important to understand the influence of the composition of the woody portion of the stem from which the RNA is extracted. This RNA from woody tissues is expected to comprise primarily of RNA from the outermost layer of woody differentiating xylem, since the final autolysis of cellular contents is slow in conifer tracheids (Bollhöner et al. [2012]) but it will also contain RNA from rays and resin duct parenchyma, as the latter can live for more than 30 years in *Pinus*(Nakaba et al. [2008]). The xylem transcriptomes obtained from the 2-year-old saplings used in this study likely reflect developmentally young, but woody, outer xylem seen immediately below the gelatinous 'developing xylem' of older trees. An expressed sequence tagged library has previously been generated from such older xylem tissues in 10-year-old *P. taeda* (Kirst et al.[2003]). Differential gene expression among samples obtained from different organs (xylem, embryo, megagametophytes) have also been reported in *P. taeda* (Loopstra and Sederoff [1995]).

Ethephon Induction of Secondary Metabolism

As expected, ethephon treatment triggered a chain of events that manifested as bark cracking and necrosis, resinification, and chlorosis, and led to significantly different transcriptomes (Table 2) in all sampled tissues. Ethephon increased the transcription of genes associated with secondary metabolism, specifically of phenylpropanoids and flavonoids in xylem tissues (Table 2). Some of these secondary metabolites may be responsible for the red-fluorescent extractives that developed in ethephon-treated xylem after fixation with FAA (Figure 4C). These extractives seem to be limited to the young outer xylem nearest to the mucilaginous xylem. This is consistent with the hypothesis that, in conifers, the ethylene signal is transported to the mucilaginous xylem which is subsequently reprogrammed to generate cells with altered physiology to produce metabolites that counter the stress factor identified by the signalling pathway (Hudgins and Franceschi [2004]). Resinification was more pronounced at the higher 5 and 25 mg ethephon dosages (data not shown); transcriptomic analyses were not attempted for these treatments due to inconsistent RNA extraction possibly due to the adverse effect of high resin content.

Ethephon and Plant Defence

Ethylene is involved in the regulation of genes involved in resistance, defence and wounding response (Van Zhong and Burns [2003]; Berrocal-Lobo and Molina [2004]; Harfouche et al.[2006]). Ethephon elicited greater transcript activity from genes implicated in defence against fungal and insect attack and oxidative stress (Table 2). Genes associated with binding (i.e. recognising) chitin, a primary component of fungal cell walls and insect exoskeletons, and chitinases, enzymes designed to degrade chitins, were up-regulated in the xylem. Ethephon exposure also led to the up-regulation of genes related to oxidative stress; active oxygen species, such as the superoxide radical generated during pathogen attack and abiotic stress, are recognised by plants as a signal for triggering defence responses (Vranova et al. [2002]).

Ethephon and Anatomical Structure of *P. Radiata*

An increase in girth within the treated area (Figure 4) was similar to that reported for balsam fir*Abies balsamea* (L.) Mill. (Eklund and Little [1996]), indicating that the effect of the ethylene generated from ethephon is limited to the area of application. The thickened cortex displayed deep fissures that did not extend to the xylem. Stereomicroscopic examination of the cut stem revealed that most of the increased growth was due to increased amounts of phloem and cortex rather than the woody portion of the stem. This confirms a previous report (Yamamoto and Kozlowski [1987a]) where a 1% ethephon dose, which is approximately equal to the 1 mg ethephon treatment in this study, was shown to elicit the most bark growth. No significant increase in diameter in the xylem portion of the stem cross sections was observed (see Figure 4) and the increased growth on the xylem side of the mucilaginous xylem consisted of traumatic parenchyma rather than xylem tissue. There was no evidence of increased lignification or increased volume in the woody cross sections. In fact, stems exposed to 1 mg ethephon exhibited an abnormal layer of unlignified tissue in the outer xylem (asterisk in Figure 4B). This corroborates the down-regulation of a cohort of genes identified via the microarray that are associated with lignin biosynthesis in the xylem which is consistent with a previous claim that ethephon reduced lignification of tracheids in compression wood that developed on the underside of tilted stems of *Pinus densiflora* (Yamamoto and Kozlowski [1987b]).

Oleoresin and Resin Canals

In response to biotic stress, conifers secrete an oleoresin that is composed mainly of mono- and diterpenes (Schmidt et al. [2010]). Oleoresin production may be associated with resin duct formation in the cambium zone and developing xylem in response to insect attack or experimentally following methyl jasmonate treatment (Zulak and Bohlmann [2010]). In this study, applications of >1 mg of ethephon induced significant resin secretion although there were no clear differences in resin duct formation (Figure 4) over the duration of the experiment. The traumatic xylem that developed in stems treated with ethephon showed poorly lignified parenchymatous cells suffused with extractives (Figure 5). The lack of change in the size and number of the resin canals indicates that the effect of ethephon did not extend to the type of resin-duct formation that is associated with terpene synthesis over the time course of this experiment. By contrast, a similar treatment (1% ethephon in lanolin) had been found previously to induce significant resinification as well as terpene-duct formation in

P. halepensis (Yamamoto and Kozlowski [1987a]). Traumatic resin-duct formation after ethephon injection has also been observed in four-year-old Douglas-fir (*Pseudotsuga menziesii* (Mirb.) Franco) and giant redwood trees (*Sequoiadendron giganteum* Lindl. Buchholz) (Hudgins and Franceschi [2004]). The higher levels of ethephon (5 and 25 mg) used in this study induced greater resin bleeding and bark splitting, and a reduction in stem girth and plant vitality leading to chlorosis.

Microarray and Quantitative RT-PCR

Heterologous hybridisation of *Pinus radiata* derived probes to the *Pinus* microarray may produce spurious hybridisation results due to imperfect annealing that can lead to incorrect interpretations of gene expression levels; quantitative RT-PCR is therefore necessary to verify the microarray results. Quantitative real-time PCR is commonly used to validate gene expression results obtained from microarray analysis however there are factors inherent in these two methods that can lead to contradictory data (Morey et al. [2006]). Microarrays are relatively insensitive in detecting<2-fold changes (Wang et al. [2006]). For example, 28% of the rice genes selected for validation gave non-significant correlations between microarray and quantitative RT-PCR signals (Dubouzet et al.[2007]), which was taken to be within normal expectations for microarray experiments. In this study, five probes were identified with consistent fold-change values by quantitative RT-PCR that confirmed the direction of expression revealed by the microarray and three of the genes had almost identical fold change estimates. Four of the probes tested could not be amplified, indicating a difference in the ability to detect target transcripts between the heterologous microarray and the PCR platforms. A greater level of probe mismatching is tolerated within a microarray hybridisation (1–6 base differences over 60-mer probe can reduce signal intensity by at least 50% (LeProust[2008]; Hughes et al. [2001])) compared with lower levels of mismatch that are permissible within a quantitative PCR experimental design. Mismatches may prevent successful RT-PCR validation with particular PCR primers or for the more dissimilar heterologous genes. In some cases, successful RT-PCR validation may be overcome through the fine tuning of PCR conditions including the re-design of an appropriate target within the microarray probe or target gene on which to base primer design so that it avoids areas of mismatch in the heterologous gene while still generating >45 bp fragments. Experimental factors such as using different reverse transcriptase-PCR kits or thermal cycling conditions may also need to be explored. It is important to note that a microarray is a tool to uncover genes of interest under particular experimental designs, and that those identified need to be confirmed using other approaches including quantitative RT-PCR. An important experimental consideration therefore is to ensure that sufficient material is archived to allow the required testing that may be required for validation.

CONCLUSIONS

It can be concluded that an annotated database of *Pinus* cDNA sequences can provide a sound basis for the design of a *Pinus* microarray to study gene expression in *P. radiata* by heterologous hybridisation. *In silico* comparisons indicated a very high similarity between publicly available transcripts derived from *P. radiata* and the microarray probes that were predominantly derived from *P. taeda*. Further genome sequencing initiatives within the Pinaceae have the potential to further expand the database upon which microarrays can be designed to interrogate transcriptome data across related species. The *Pinus* microarray was tested by heterologous hybridisation with cRNA produced from bark, mucilaginous xylem and xylem of *P. radiata* clonal saplings that were subjected to an ethephon induction trial. The microarray was able to highlight that ethephon initiated very different transcriptomes for all tissues sampled at 1 and 8 WAT and these could be associated with an expected cascade of physiological responses. As is common practice with microarray experiments validation of gene expression highlighted by the *Pinus* microarray via quantitative RT-PCR is required; a caveat for validation of heterologous hybridisation is that primer design needs to overcome sequence differences that can exist between the heterologous test probes and the microarray 60-mer sequences. While not further tested here, the *Pinus* microarray developed may be useful for *P. taeda* and other closely related species in the Pinaceae since the majority (>98%) of sequences originate from *P. taeda*. Further development of the All-*Pinus* gene database is also valuable as more sequence data from related species in the Pinaceae becomes available.

AUTHORS' CONTRIBUTIONS

All authors contributed to this paper. JDG contributed initial idea and concept development, microarray design, experimental manipulations and manuscript writing; LD provided microscopy and interpretation; MAB and LM implemented microarray hybridisation and analysis; VL contributed database construction; and GLl-J contributed to concept development and manuscript writing. All authors read and approved the final manuscript.

ACKNOWLEDGMENTS

The authors wish to express their gratitude to the following Scion staff: Lucy Macdonald, Lisa Stanbra, Lorelle Phillips and, most specially, Heather Flint for their critical support in various technical aspects of this research. Drs. Tim Strabala and Armin Wagner provided invaluable insight and discussions that vastly improved the manuscript. Details of the microarray experiment have been submitted according to the specifications of the National Centre for Biotechnology Information (NCBI) Gene Expression Omnibus (GEO) and are available athttp://www.ncbi.nlm.nih.gov/geo/query/acc.cgi?acc=GSE42300.

REFERENCES

1. Adomas, A, Heller, G, Li, G, Olson, A, Chu, T-M, Osborne, J, Craig, D, van Zyl, L, Wolfinger, R, Sederoff, R, Dean, RA, Stenlid, J, Finlay, R, Asiegbu, FO (2007). Transcript profiling of a conifer pathosystem: response of *Pinus sylvestris* root tissues to pathogen (*Heterobasidion annosum*) invasion. *Tree Physiology, 27*(10), 1441–1458.
2. Altschul, SF, Madden, TL, Schäffer, AA, Zhang, J, Zhang, Z, Miller, W, Lipman, DJ (1997). Gapped BLAST and PSI-BLAST: a new generation of protein database search programs. *Nucleic Acids Research, 25*(17), 3389–3402.
3. Balk, P, Bronnum, P, Perks, M, Stattin, E, van der Geest, L, van Wordragen, M (2007). Innovative Cold Tolerance Test for Conifer Seedlings. National Proceedings: Forest and Conservation

Nursery Associations-2006. RMRS-P-50. (pp. 9–14). USDA Forest Service, Rocky Mountain Research Station, Fort Collins, CO.

4. Benjamini, Y, & Hochberg, Y (1995). Controlling the false discovery rate: a practical and powerful approach to multiple testing. *Journal of the Royal Statistical Society. Series B (Methodological)*,*57*(1), 289–300.
5. Berrocal-Lobo, M, & Molina, A (2004). Ethylene response factor 1 mediates Arabidopsis resistance to the soilborne fungus *Fusarium oxysporum*. *Molecular Plant-Microbe Interactions, 17*(7), 763–770.
6. Bollhöner, B, Prestele, J, Tuominen, H (2012). Xylem cell death: emerging understanding of regulation and function. *Journal of Experimental Botany, 63*(3), 1081–1094.
7. Brown, K, & Leopold, A (1973). Ethylene and the regulation of growth in pine. *Canadian Journal of Forest Research, 3*(1), 143–145.
8. Cato, S, McMillan, L, Donaldson, L, Richardson, T, Echt, C, Gardner, R (2006). Wood formation from the base to the crown in *Pinus radiata*: gradients of tracheid wall thickness, wood density, radial growth rate and gene expression. *Plant Molecular Biology, 60*(4), 565–581.
9. Davey, M, Graham, N, Vanholme, B, Swennen, R, May, S, Keulemans, J (2009). Heterologous oligonucleotide microarrays for transcriptomics in a non-model species; a proof-of-concept study of drought stress in *Musa*. *BMC Genomics, 10*(1), 436. | BioMed Central Full Text
10. Dubouzet, JG, Ishihara, A, Matsuda, F, Miyagawa, H, Iwata, H, Wakasa, K (2007). Integrated metabolomic and transcriptomic analyses of high-tryptophan rice expressing a mutant anthranilate synthase alpha subunit. *Journal of Experimental Botany, 58*(12), 3309–3321.
11. Eklund, L, & Little, CHA (1996). Laterally applied Ethrel causes local increases in radial growth and indole-3-acetic acid concentration in *Abies balsamea* shoots. *Tree Physiology, 16*(5), 509–513.
12. Feldmesser, E, Rosenwasser, S, Vardi, A, Ben-Dor, S (2013). Improving automated de-novo transcriptome definition in

non-model organisms by integrating manually defined gene information.*EMBnet. Journal*, *19*(A), 45–46.

13. Hall, TA (1999). BioEdit: a user-friendly biological sequence alignment editor and analysis program for Windows 95/98/NT. *Nuclear Acids Symposium Series*, *41*, 95–98.
14. Harfouche, AL, Shivaji, R, Stocker, R, Williams, PW, Luthe, DS (2006). Ethylene signaling mediates a maize defense response to insect herbivory. *Molecular Plant Microbe Interactions: MPMI*, *19*(2), 189–199.
15. Heath, LS, Ramakrishnan, N, Sederoff, RR, Whetten, RW, Chevone, BI, Struble, CA, Jouenne, VY, Chen, D, van Zyl, L, Grene, R (2002). Studying the functional genomics of stress responses in loblolly pine with the Expresso microarray experiment management system. *Comparative and Functional Genomics*, *3*(3), 226–243.
16. Hiremath, PJ, Farmer, A, Cannon, SB, Woodward, J, Kudapa, H, Tuteja, R, Kumar, A, Bhanuprakash, A, Mulaosmanovic, B, Gujaria, N, Krishnamurthy, L, Gaur, PM, Kavikishor, PB, Shah, T, Srinivasan, R, Lohse, M, Xiao, Y, Town, CD, Cook, DR, May, GD, Varshney, RK (2011). Large-scale transcriptome analysis in chickpea (*Cicer arietinum* L.), an orphan legume crop of the semi-arid tropics of Asia and Africa. *Plant Biotechnology Journal*, *9*(8), 922–931.
17. Huang, X, & Madan, A (1999). CAP3: A DNA sequence assembly program. *Genome Research*, *9*(9), 868.
18. Huang, DW, Sherman, BT, Lempicki, RA (2008). Systematic and integrative analysis of large gene lists using DAVID bioinformatics resources. *Nature Protocols*, *4*(1), 44–57.
19. Huang, DW, Sherman, BT, Lempicki, RA (2009). Bioinformatics enrichment tools: paths toward the comprehensive functional analysis of large gene lists. *Nucleic Acids Research*, *37*(1), 1–13.
20. Huang, Y-L, Fang, X-T, Lu, L, Yan, Y-B, Chen, S-F, Hu, L, Zhu, C-C, Ge, X-J, Shi, S-H (2012). Transcriptome analysis of an invasive weed *Mikania micrantha*. *Biologia Plantarum*, *56*(1), 111–116.
21. Hudgins, J, & Franceschi, V (2004). Methyl jasmonate-induced ethylene production is responsible for conifer phloem defense responses and reprogramming of stem cambial zone for traumatic

resin duct formation. *Plant Physiology, 135*(4), 2134.

22. Hughes, T, Mao, M, Jones, A, Burchard, J, Marton, M, Shannon, K, Lefkowitz, SM, Ziman, M, Schelter, JM, Meyer, MR, Kobayashi, S, Davis, C, Dai, H, He, YD, Stephaniants, SB, Cavet, G, Walker, WL, West, A, Coffey, E, Shoemaker, DD, Stoughton, R, Blanchard, AP, Friend, SH, Linsley, PS (2001). Expression profiling using microarrays fabricated by an ink-jet oligonucleotide synthesizer. *Nature Biotechnology, 19*(4), 342–347.
23. Kirst, M, Johnson, A, Baucom, C, Ulrich, E, Hubbard, K, Staggs, R, Paule, C, Retzel, E, Whetten, R, Sederoff, R (2003). Apparent homology of expressed genes from wood-forming tissues of loblolly pine (*Pinus taeda* L.) with *Arabidopsis thaliana. Proceedings of the National Academy of Sciences,100*(12), 7383–7388.
24. [http://www.agilent.com/chem/dna] LeProust, E. (2008). Agilent's Microarray Platform: How High-Fidelity DNA Synthesis Maximizes the Dynamic Range of Gene Expression Measurements Agilent's Application Notes (p. 12), Agilent Technologies ().
25. Li, X, Wu, HX, Southerton, SG (2012). Identification of putative candidate genes for juvenile wood density in *Pinus radiata. Tree Physiology, 32*(8), 1046–1057.
26. Lieberman, M (1979). Biosynthesis and action of ethylene. *Annual Review of Plant Physiology,30*(1), 533–591.
27. Loopstra, CA, & Sederoff, RR (1995). Xylem-specific gene expression in loblolly pine. *Plant Molecular Biology, 27*(2), 277–291.
28. Lorenz, WW, Alba, R, Yu, YS, Bordeaux, J, Simoes, M, Dean, J (2011). Microarray analysis and scale-free gene networks identify candidate regulators in drought-stressed roots of loblolly pine (*P. taeda* L.). *BMC Genomics, 12*(1), 264. | BioMed Central Full Text
29. Lorenz, WW, Ayyampalayam, S, Bordeaux, JM, Howe, GT, Jermstad, KD, Neale, DB, Rogers, DL, Dean, JFD (2012). Conifer DBMagic: a database housing multiple *de novo* transcriptome assemblies for 12 diverse conifer species. *Tree Genetics Genomics, 8*, 1477–1485.
30. Min, XJ, Butler, G, Storms, R, & Tsang, A. (2009). Comparative Assessment of DNA Assemblers for Assembling Expressed

Sequence Tags. In Bioinformatics (OCCBIO'09. Ohio Collaborative Conference on, 2009, pp. 79–82). IEEE. doi:10.1109/OCCBIO.2009.19.

31. Morey, JS, Ryan, JC, Van Dolah, FM (2006). Microarray validation: factors influencing correlation between oligonucleotide microarrays and real-time PCR. *Biological Procedures Online, 8*(1), 175–193.
32. Myburg, H, Morse, A, Amerson, H, Kubisiak, T, Huber, D, Osborne, J, Osborne, JA, Garcia, SA, Nelson, CD, Davis, JM, Covert, SF, van Zyle, LM (2006). Differential gene expression in loblolly pine (*Pinus taeda* L.) challenged with the fusiform rust fungus, *Cronartium quercuum f. sp. fusiforme.Physiological and Molecular Plant Pathology, 68*(1–3), 79–91.
33. Nakaba, S, Kubo, T, Funada, R (2008). Differences in patterns of cell death between ray parenchyma cells and ray tracheids in the conifers *Pinus densiflora* and *Pinus rigida. Trees-Structure and Function, 22*(5), 623–630.
34. New Zealand Forest Owners Association. (2011). New Zealand Plantation Forest Industry: Facts and Figures. In N. Z. F. O. Association (Ed.), (Vol. 2010/11, p. 4). New Zealand: New Zealand Forest Owners Association.
35. Perdiguero, P, Collada, C, Barbero, MC, Casado, GG, Cervera, MT, Soto, Á (2012). Identification of water stress genes in *Pinus pinaster* Ait. by controlled progressive stress and suppression-subtractive hybridization. *Plant Physiology and Biochemistry, 50*(1), 44–53.
36. Pfaffl, MW (2004). Quantification strategies in real-time PCR. *AZ of Quantitative PCR, 1*, 89–113.
37. Reinders, J, Vivier, CD, Theiler, G, Chollet, D, Descombes, P, Paszkowski, J (2008). Genome-wide, high-resolution DNA methylation profiling using bisulfite-mediated cytosine conversion. *Genome Research, 18*(3), 469–476.
38. Schmidt, A, Wächtler, B, Temp, U, Krekling, T, Séguin, A, Gershenzon, J (2010). A bifunctional geranyl and geranylgeranyl diphosphate synthase is involved in terpene oleoresin formation in*Picea abies*. *Plant Physiology, 152*(2), 639–655.
39. Schnurr, JP, Cheng, Z, Boe, AA (1996). Effects of plant growth

regulators on sturdiness of Jack pine seedlings. *Journal of Environmental Horticulture*, *14*, 228–230.

40. Smyth, GK In Gentleman R, Carey VJ, Huber W, Irizarry RA, Dudoit S (Eds.) (2005). Limma: Linear Models for Microarray Data. *'Bioinformatics and Computational Biology Solutions Using R and Bioconductor'* (pp. 397–420). Springer Science+Business Media New York, New York.
41. Sun, WQ, & Bassuk, NL (1993). Auxin-induced Ethylene Synthesis during Rooting and Inhibition of Budbreak ofRoyalty'Rose Cuttings. *Journal of the American Society for Horticultural Science*,*118*(5), 638–643.
42. Van Zhong, G, & Burns, JK (2003). Profiling ethylene-regulated gene expression in *Arabidopsis thaliana* by microarray analysis. *Plant Molecular Biology*, *53*(1), 117–131.
43. Van Zyl, L, Von Arnold, S, Bozhkov, P, Chen, Y, Egertsdotter, U, MacKay, J, Sederoff, RR, Shen, J, Zelena, L, Clapham, DH (2002). Heterologous array analysis in Pinaceae: hybridization of *Pinus taeda* cDNA arrays with cDNA from needles and embryogenic cultures of *P. taeda, P. sylvestris* or*Picea abies*. *Comparative and Functional Genomics*, *3*(4), 306–318.
44. Vranova, E, Inzé, D, Van Breusegem, F (2002). Signal transduction during oxidative stress. *Journal of Experimental Botany*, *53*(372), 1227–1236.
45. Wang, Y, Barbacioru, C, Hyland, F, Xiao, W, Hunkapiller, KL, Blake, J, Chan, F, Gonzalez, C, Zhang, L, Samaha, RR (2006). Large scale real-time PCR validation on gene expression measurements from two commercial long-oligonucleotide microarrays. *BMC Genomics*, *7*(1), 59.
46. Warner, H, & Leopold, A (1969). Ethylene evolution from 2-chloroethylphosphonic acid. *Plant Physiology*, *44*(1), 156.
47. Yamamoto, F, & Kozlowski, T (1987). Effect of ethrel on growth and stem anatomy of *Pinus halepensis* seedlings. *IAWA Bulletin*, *8*(1), 11–19.
48. Yamamoto, F, & Kozlowski, T (1987). Effects of flooding, tilting of stems, and ethrel application on growth, stem anatomy and ethylene production of *Pinus densiflora* seedlings. *Journal of Experimental Botany*, *38*(2), 293–310.

49. Yang, S-H, van Zyl, L, No, E-G, Loopstra, CA (2004). Microarray analysis of genes preferentially expressed in differentiating xylem of loblolly pine (*Pinus taeda*). *Plant Science, 166*(5), 1185–1195.
50. Zulak, KG, & Bohlmann, J (2010). Terpenoid biosynthesis and specialized vascular cells of conifer defense. *Journal of Integrative Plant Biology, 52*(1), 86–97.

Chapter 8

Isolation and Expression Analysis of CBF4 from Vitis Amurensis Associated with Stress

Chang Dong[1,2], Meng Zhang[1], Zhiying Yu[1], Junpeng Ren[1], Yang Qin[2], Bailin Wang[2], Lizhen Xiao[2], Zhen Zhang[1], and Jianmin Tao[1]

[1]College of Horticulture, Nanjing Agricultural University, Nanjing, China

[2]Department of Horticulture, Heilongjiang Academy of Agricultural Science, Harbin, China

ABSTRACT

The C-repeat Binding Factor/Dehydration Responsive Element Binding factor (CBF/DREB) responsive pathway is involved in plant response to abiotic stress. In this study, we report the isolation of VaCBF4 (a complete cDNA) and its promoter from Vitis amurensis through rapid amplification of cDNA ends and genome walking techniques, respectively. The CBF4 transcript accumulation of V. amurensis increased under cold, salinity, and abscisic acid (ABA) and salicylic acid (SA) treatments, whereas that of Vitis vinifera showed a different change. The transcript levels of VaCBF4 in the roots, stems, leaves, and petioles under cold, salinity, and ABA and SA treatments were up-regulated in a timedependent manner. The presence of the cis-elements MBC, ABRE, and as-2-box in the VaCBF4 promoter further confirmed that this promoter is a component of the CBF transduction pathway, which is involved in plant response to multistress.

INTRODUCTION

Low temperature, high salinity, and drought are environmental factors that limit the geographical distribution and affect the growth, development, and productivity of plants. Plants respond and adapt to these stresses at the molecular, physiological, and biochemical levels. The expression levels of various genes in plants are activated by these stresses [1, 2]. The products of these genes regulate the gene expression and signal transduction in response to stress. The cis-elements involved in stress response were analyzed to elucidate the molecular mechanisms of gene expression in response to stress. The C-repeat/Dehydration Responsive Element (CRT/DRE) with the core CCGAC sequence is a cis-acting element in the promoter that regulates gene expression in response to cold and drought in Arabidopsis [3, 4]. A similar motif was identified as the Low-Temperature Responsive Element (LTRE) in cold-inducible genes [5].

At present, the most understood genetic system with an important role in cold acclimation is the Arabidopsis CBF (CRT/DRE-binding transcription factors) cold response pathway [6]. Exposing Arabidopsis plants to low temperature or drought results in the rapid induction of a small family of transcriptional factors known as CBF1, CBF2, CBF3(or as DREB1B, DREB1C, and DREB1A, respectively) [7], and CBF4 [8]. CBF/DREB-type proteins belong to the AP2/ERF (APETALA 2/Ethylene Response Factors) DNA-binding protein family. Moreover, the constitutively over-expression of CBF genes induces the expression of downstream genes and improves the freezing and salt tolerance of transgenic plants [9].

Despite the high conservation of CBF residues in plants, some structural and regulatory differences were observed in the CBF cold response pathway among plant species. For example, the CBF regulon of dwarf apple was found to be induced more than that of Arabidopsis [10]. A previous work on blueberry showed that CBF genes from different species have dramatically different levels of induction under low temperature [11]. Although recent breakthroughs have increased the knowledge on the molecular basis of the CBF cold pathway in herbaceous species, little is known in woody plants. To date, data on CBF genes from perennial plants include functional analysis of CBF from cherry [12], eucalyptus [13], blueberry [11], and apple [10].

Grape is one of the most important fruit crops in the world, and its production has been expanding as more varieties and better quality wines become available. However, low temperature limits the geographical distribution and affects the growth, development, and productivity of this crop. Vitis amurensis is a freezing-tolerant wild grape species native to Northern China. This species can survive below 40°C [14]. V. amurensis is one of the most widely used species for rootstock and winemaking in grape cultivation. However, little is known about the freezing tolerance of V. amurensis at the molecular level.

In this study, the VaCBF4 gene was isolated from V. amurensis. We also compared VaCBF4 gene expression in the leaves of cold-tolerant and cold-sensitive grapes with that in tissues of V. amurensis at different times of low temperature, high salinity, and abscisic acid (ABA) and salicylic acid (SA) treatments, and then confirmed its transcript patterns under stress.

MATERIAL AND TREATMENTS

Sixty-day-old in vitro samples of wild V. amurensis and Vitis vinifera' Manicure Finger' grown in Murashige and Skoog (MS) medium under a 16 h light/8 h dark regime at 25°C were subjected to the following treatments for stress-responsive gene expression: cold stress by transferring to 4°C, salinity stress by supplying 200 mM NaCl, and hormonal treatments by supplying ABA (10 μM) or SA (5 μM) directly to the plants at different times.

Isolation of VaCBF4

Total RNA from non-stressed or cold-stressed wild samples in vitro was extracted using the CTAB method (Lin et al., 2008) and converted to cDNA using Primer Mix according to the instructions of Rever Tra Ace RT Kit (Toyobo, Japan). The partial coding sequence of the CBF4 genes was amplified using gene-specific primers, C4-F and C4-R (Table 1). The primers were designed based on the genome of Vitis (http://www.vitisgenome.it). The 3′ regions of the corresponding cDNA fragments were obtained by 3′ rapid amplification of cDNA ends (RACE) using the cDNA amplification kit (Clontech, CA, USA) with the gene-specific primers 16326 and C4-rF2 for the first step and 16324 and C4-rF3 for the second step (Table 1). Amplification was carried out under the following conditions: 94°C for 4 min; 36 cycles of 94°C for 30 s, 60°C for 45 s, 64°C for 60 s, and 72°C for 60 s; and a final extension at 72°C for 10 min.

The isolation of the promoter sequence was carried out using a Universal Genome Walker kit (UGWK, Clontech, USA) on 10 ng of V. amurensis genomic DNA digested by four blunt-end-generating restriction enzymes (EcoR V, DraI, SacI, and StuI). After purification, the restriction fragments were ligated with Genome Walker adaptors (Ad1 and Ad2, Table 1). The polymerase chain reaction (PCR) products were amplified on each restriction fragment set using the primers C4-R1 and ada1 for the first step PCR and C4-R2 and ada 2 for the second step PCR (Table 1). The PCR products were cloned and sequenced by the pMD19-T vector.

Sequence Analysis

Homology searches were performed using BLAST with default parameters at NCBI (http://blast.ncbi.nlm.nih.gov/Blast.cgi). Multiple sequence alignment of CBFs was performed using Crustal 1.83, and shading was carried out using DNAMAN software (Version 5.2.2, Lynnon Biosoft). The CBF protein sequences necessary for constructing a phylogenetic tree were retrieved from the GenBank database and depicted using Crustal 1.83. Phylogenetic analysis was based on the neighbor-joining method using Mega 4.1 software with 1000 bootstrap replicates and substitution model of Poisson correction. Promoter sequence analysis was performed using PLACE (http://www.dna.affrc.go.jp) and Plant CARE.

Real-Time Reverse Transcription Quantitative-PCR (RT-qPCR)

Total RNA from different treatments and tissues were digested with RNase-free DNAase I (TaKaRa, Japan) to remove genomic DNA contamination. cDNA fragments were synthesized as aforementioned. Transcript level was determined on an ABI PRISM 7300 (Applied Biosystems, Foster City, CA, USA). For each reaction, 20 ng of cDNA was used with SYBR Premix Ex Taq mix (TaKaRa, Japan), including primers C4rt1 and C4rt2 for the CBF4 gene. All experiments were run in triplicates with different synthesizedc DNA fragments. A positive control was provided through a parallel analysis using the grape malate dehydrogenase gene as an internal reference [15], and was calculated relative to a calibrator using the formula $2^{-\Delta\Delta Ct}$.

RESULTS

Identification of the VaCBF4 Gene from V. Amurensis

We constructed a cDNA admixture template from the cold-treated and non-treated in vitro leaves to identify the gene related to low temperature. A CBF4-like partial sequence was found, suggesting that the corresponding gene was probably induced by low temperature. Given that this sequence contained the intact 5′ end of the Open Reading Frame (ORF) of a putative CBF4 gene, we can obtain the rest of the sequence by 3′ RACE assay. A fragment of 513 bp was amplified through nested PCR. Based on the sequence analysis, we obtained a cDNA sequence designated as VaCBF4 (GenBank accession No. HM231149). VaCBF4 contains an ORF of 654 bp, a 5′ untranslated region (UTR) of 22 bp, and a 3′ UTR of 278 bp. Moreover, this cDNA contains a single ORF of 218 amino acids and encodes a putative protein with a predicted molecular mass of 24.23 kDa and a pI of 5.4.

The obtained amino acid encodes an AP2/ERF DNA binding protein (Figure 1). We found putative nuclear localization signals (NLS) from the N-terminus to the C-terminus, as well as the domains AP2, DSAWR, and LWSY in the C-terminal region.

Table 1: Primers used in this study

Primer	Sequence (5′ – 3′)
C4-F	CGACACTTCAGTCTTCACCGTT
C4-R	ACTGAACTCCCACCTACCCTCT
C4-rF2	GGAGGAAGAAGTTTCGGGAGAC
16326	GGTGGTAGAGCTCGCAGGACTGCAGCTGACTG
C4-rF3	CAAGATGAGAGGAGGGAAGAGAGTGA
16324	AGAGCTCGCAGGACTGCAGCTGACTGACTAC
Ad1	GATTTACGACTCACTATAGGGCACGCGTGGTCGACGGCCCGGGCTGGT
Ad2	PO4-ACCAGCCC-NH2
C4-R1	CCCACCTACCCTCTGTTACTTT
ada1	GATTTACGACTCACTATAGGGC
C4-R2	TCACTCTCTTCCCTCCTCTCATCTTG
ada2	ACTATAGGGCACGCGTGGT
C4rt1	GACACTTCAGCCTTCACCG
C4rt2	CCTCGCATACCCACTTCC

Phylogenetic Analysis of VaCBF4

A BLAST search was performed using the deduced amino acid sequence of VaCBF4. Several homologous sequences of VaCBF4 in higher plants were found in the GenBank database. Alignment analysis and domain comparison indicated that VaCBF4 shared high homology to other proteins in the AP2/ERF domain (Figure 1). Through systematic phylogenetic analysis, AP2/ERF proteins were classified into 15 plant CBF proteins using the neighbor-joining phylogenetic method (Figure 2). The results showed that the sequence from V. amurensis belong to the CBF4 cluster of CBF proteins.

Sequence Analysis of the VaCBF4 Promoter

The VaCBF4 promoter was obtained using the UGWK method (Figure 2). The result revealed that a typical TATA-box was located at −108 bp upstream of the ATG translational initiation codon. Furthermore, cis-acting elements involved in stress-related responses were found in the VaCBF4 promoter. These elements, which include ACE, as-2-box, MeJA-like, MYB, ARE, MYC (MYC recognition site), LTRE, and ABRE, are necessary for stress responsive expression under abiotic stress.

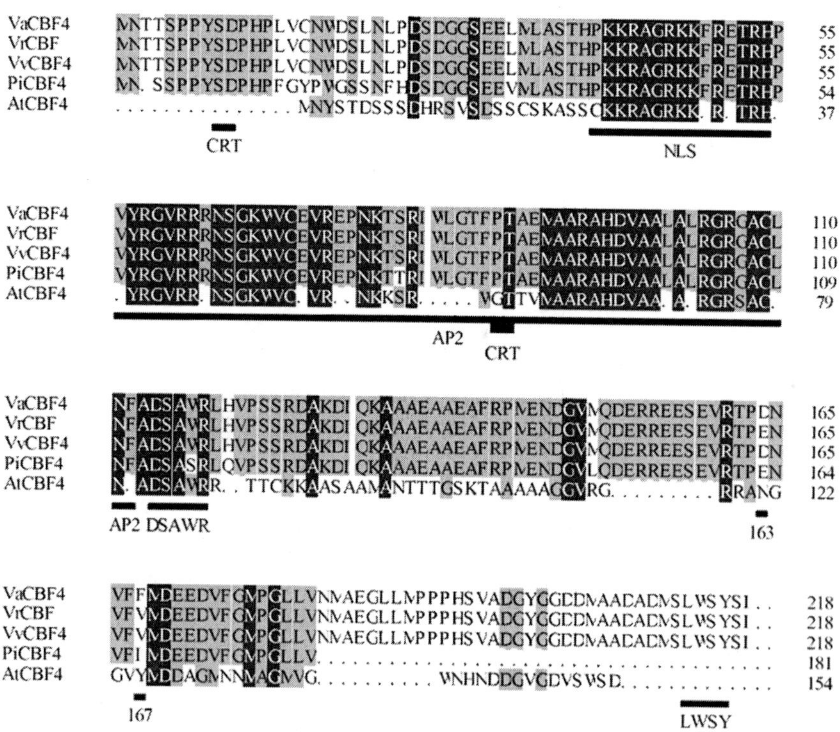

Figure 1: Multiple alignment of VaCBF4 with the CBF4 proteins of Vitaceae, Parthenocissus, and Arabidopsis. The amino acid sequences are from NCBI [AtCBF4 (np_200012), PiCBF4 (ABU55674), VrCBF4 (AY706986), and VvCBF4 (XP_00228 0097)]. The nuclear localization signal, AP2, DSAWR, and LWSY domains are indicated by bars.

Differential Accumulation of Vitis CBF4 Under Low Temperature Stress

The CBF4 gene expression in the leaves was compared in vitro with that in the relatively cold-hardy wild V. amurensis and the relatively cold-sensitive V. vinifera cultivar "Manicure Finger" (Figure 3(a)). V. amurens reached its maximum midwinter cold hardiness level at <−40°C, whereas "Manicure Finger" was cold hardy at approximately −7°C. Under cold treatment, the expression of CBF4 in V. amurensis was not substantially different from that in V. vinifera. However, a significant difference in transcription was found between the two species under natural conditions. Unlike the transcript accumulation of VvCBF4, the transcript accumulation of VaCBF4 in the leaves, roots, stems, and petioles was not observed under natural conditions. This finding was further confirmed by quantitative RT-PCR (Figure 3(b)).

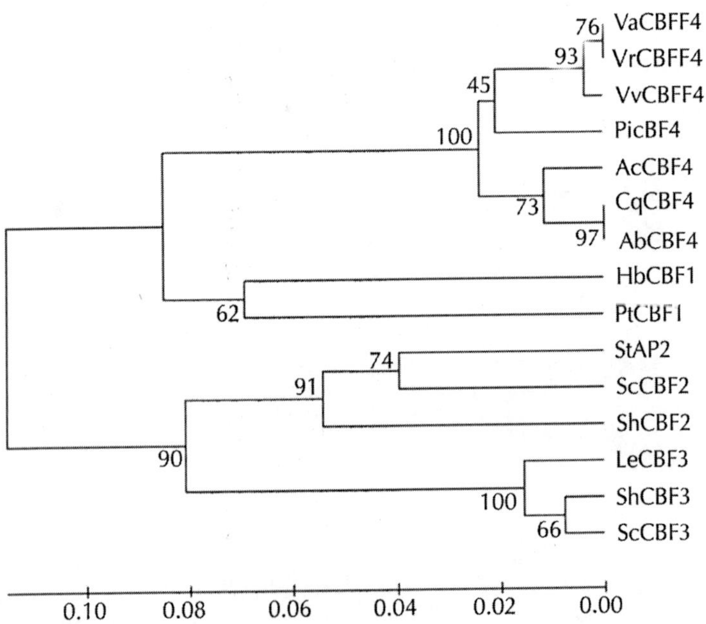

Figure 2: Bootstrap phylogenetic analysis of the selection of plant CBF proteins from NCBI. The amino acid sequences are VvCBF4 (XP_002280097), AbCBF4 (ABU55675), VrCBF (AAW58104), PiCBF4 (ABU55674), AcCBF4 (ABU55676), CqCBF4 (EF150912), HbCBF1 (AAY43213), PtCBF1

(ABO48363), ScCBF2 (ACB45094), ShCBF2 (ACB45080), ShCBF3 (ACB45078), and ScCBF3 (ACB45092).

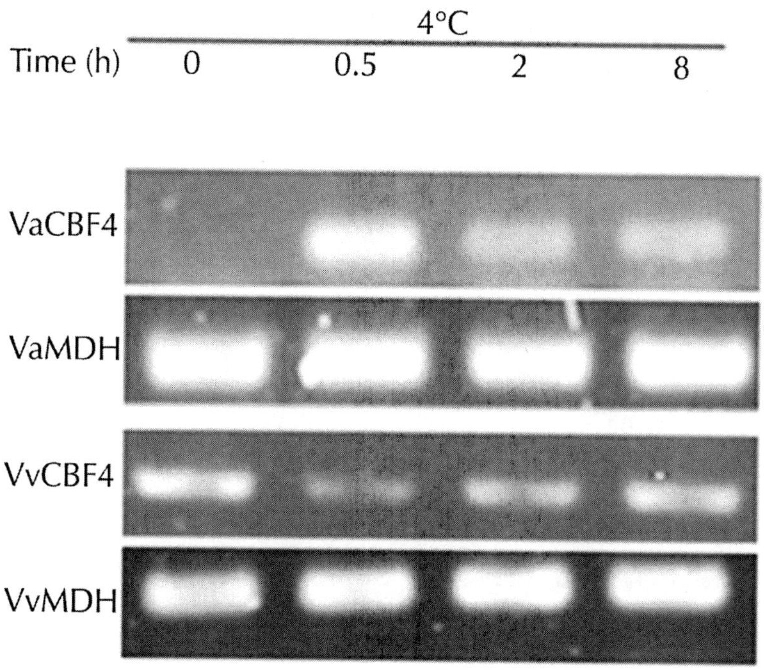

(a)

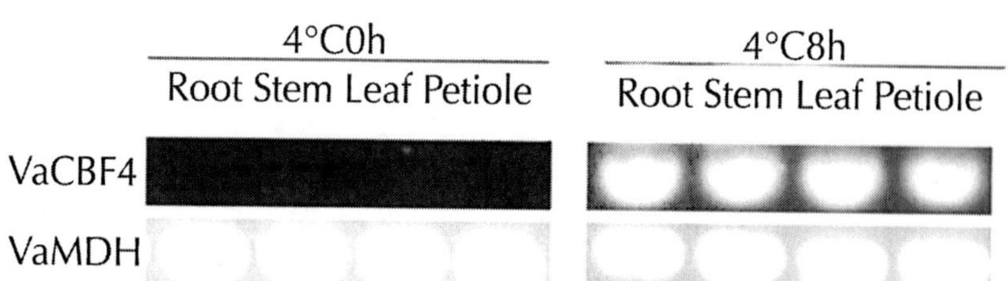

(b)

Figure 3: Transcript accumulation of Vitis CBF4 under natural and low temperature stresses. (a) An expression difference of CBF4 in the cold-hardy V.

amurensis (Va) and the less cold-hardy V. vinifera "Manicure Finger" (Vv); (b) Expression of organism in V. amurensis.

Accumulations of each transcript in the roots, stems, leaves, and petioles were analyzed by quantitative RTPCR to further determine the acquisition mechanisms of cold acclimation in relation to VaCBF4 gene expression. As shown in Figure 4, VaCBF4 exhibited different expression patterns in the organs under low temperature treatment. Accumulations of VaCBF4 transcript in all organs were inactivated by low temperature prior to treatment at 0.5 h compared with the corresponding controls. During the subsequent chilling period, the transcript level of VaCBF4 in the stems and petioles slightly increased at the early stage of treatment and then reached the peak on day 2 or day 3. By contrast, the transcript level of VaCBF4 in the roots sharply reached the peak at 4 h and then gradually decreased thereafter. However, the mRNA expression in the leaves was slightly detected at 0.5 h and then remained almost constant after over 3 d of treatment at 4°C.

Expression Pattern of VaCBF4 in Salinity Stress

We analyzed the expression of the VaCBF4 gene under high salinity stress to obtain more information about its expression under stress. The results showed that VaCBF4 expression was induced by salinity throughout the treatment period (Figure 5). However, distinct organs displayed marked differences in gene expression levels during the treatment period. The expression of VaCBF4 in the stems and leaves reached the peak after approximately 2 d of treatment. The peak transcript accumulation in the roots was observed at 6 h, whereas that in the petioles was wavy.

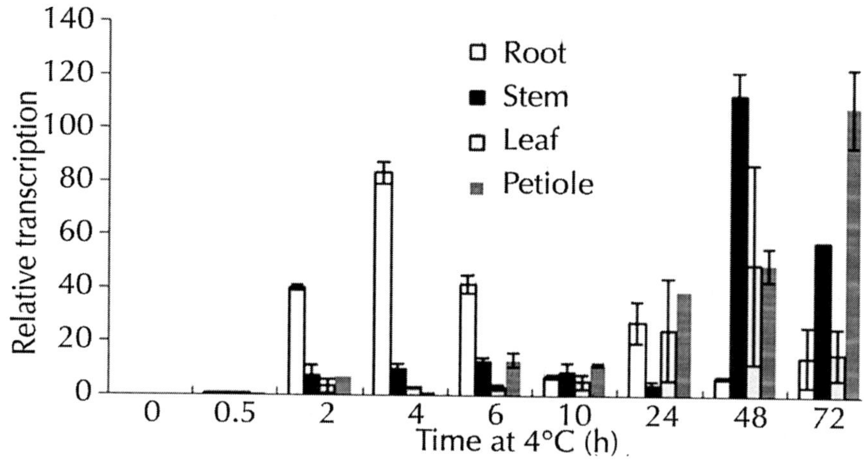

Figure 4: Changes in the mRNA accumulation of VaCBF4 in the roots, stems, leaves, and petioles under 4°C treatment.

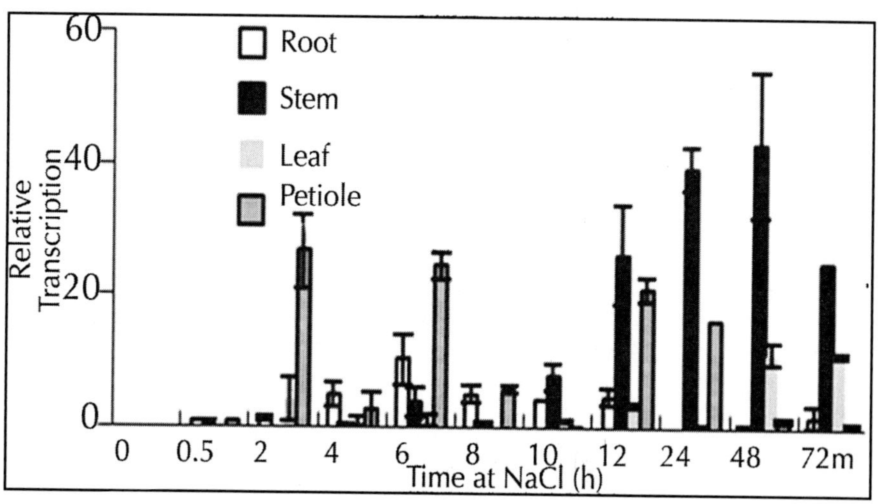

Figure 5: Changes in the mRNA accumulation of VaCBF4 in the roots, stems, leaves, and petioles during under 200 mM NaCl treatment.

ABA and SA are Regulatory Factors for VaCBF4

ABA and SA have important functions in physiological and genetic regulation during stress. We hypothesized that the expression pattern of VaCBF4 may be affected by ABA and SA. Therefore, we measured the transcript level of the organs treated with ABA and SA, and compared their expression patterns. The expression of VaCBF4 was induced by exogenous ABA and SA treatments in Figure 6. However, the organs displayed significant differences in gene expression levels. Under ABA and SA treatments, the peak pattern in the roots was found at the early stage, whereas that in the stems was observed at the later stage. However, the expression pattern of VaCBF4 in the leaves under ABA and SA treatments was wavy. The peak VaCBF4 expression in the leaves under ABA treatment appeared at 1, 8, and 24 h, whereas that in the leaves under SA treatment appeared at 2, 6, and 24 h.

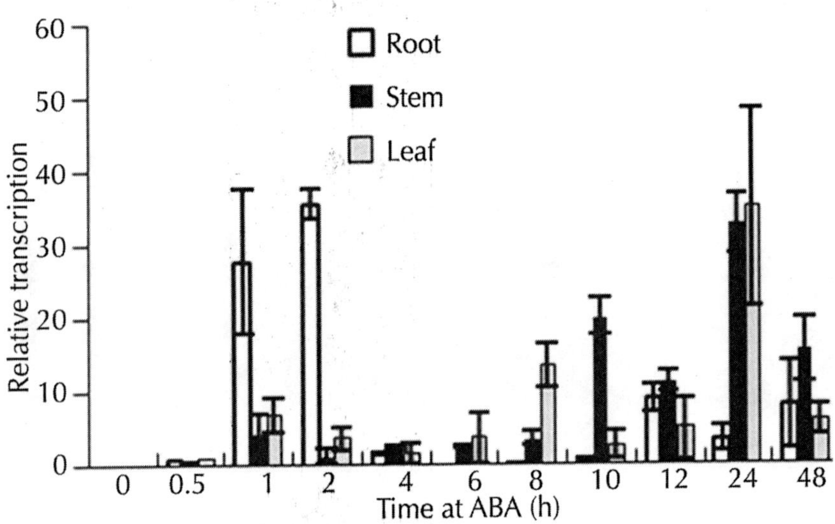

(a)

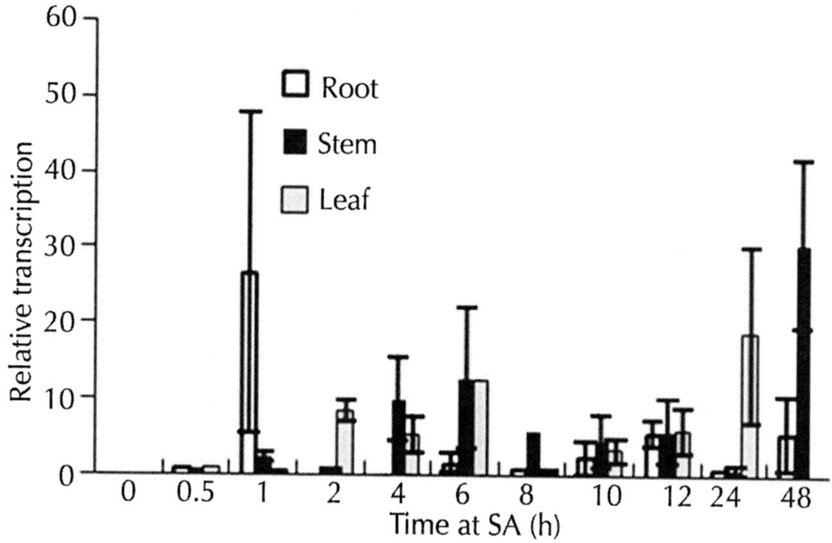

(b)

Figure 6: Changes in the mRNA accumulation of VaCBF4 in the roots, stems, and leaves under ABA and SA treatments.

DISCUSSION

We isolated and characterized a complete cDNA sequence of CBF4 from V. amurensis. The N-terminal region of VaCBF4 was found to share the conserved CBF4 domains with that of the other known CBF4 (Figure 1). Vitis CBF4 was proven to be a nuclear protein [15]. Moreover, Va CBF4 was able to map to the CBF4 group in the CBF/DREB subfamily through phylogenetic analysis. These data indicated that VaCBF4 is a CBF domain protein that may function as a transcription factor in V. amurensis.

Cold induces the expression of most reported CBF genes, including VaCBF4. VaCBF4 transcripts were detectable in different tissues (roots, stems, leaves, and petioles) under low temperature treatment. These transcripts remain present for a long period, with lower expression in the leaves than in the other tissues. A similarly cold-induced gene expression was reported for the CBF4 of Vitis riparia and V. vinifera [15]. In agreement with the high identity between their amino acid domains, the expression patterns were almost the same between V. amurensis and V. vinifera under low temperature treatment. Interestingly, under natural conditions, the CBF4 expression of V. vinifera was visible in the leaves [15], whereas that of V. amurensis was not detectable in the leaves, stems, roots, and petioles. This finding was further confirmed by quantitative RT-PCR. The important role of CBF4 proteins in the freezing tolerance of V. amurensis may be attributed to the difference in regulon size between V. vinifera and V. amurensis, given the high similarity between VaCBF4 and VvCBF4 (only substitution at position 167 of one negatively charged amino acid, F, for another, V). Haake et al. [8] showed that Arabidopsis CBF4 expression is induced by drought and ABA treatments, but not by cold treatment, under experimental conditions. The freezing-hardy materials may have more CBF4 regulons than the freezing-sensitive materials. However, this assumption needs further investigation. In addition, we showed that the accumulation of CBF4 transcripts from V. amurensis was dramatically induced by salinity and exogenous ABA and SA treatments. Hence, we propose that the CBF4 gene also functions in the signal transduction of high salinity and hormone adaption in V. amurensis. However, the relative transcript levels of VaCBF4 in different tissues exhibited different changes after exposure to different stress treatments.

An analysis of regulatory elements in the upstream of the VaCBF4 gene by PLACE and Plant CARE databases covered the putative DNA-binding sites (Figure 7). The promoter sequences of this gene were found to contain several cis-regulatory elements, such as LTRE, ABRE, MYC (MYC recognition site), and MBS (MYB bind site). These elements are related to stress response, specifically to cold/dehydration. These cis-regulatory elements are conserved in several plant species. The presence of these conserved motifs confirmed that the VaCBF4 gene was regulated by cold, high salinity, and exogenous ABA and SA treatments. The multiple stressresponse of the VaCBF4 gene also suggests that V. amurensis should be used for further research on molecular mechanism and stress breeding.

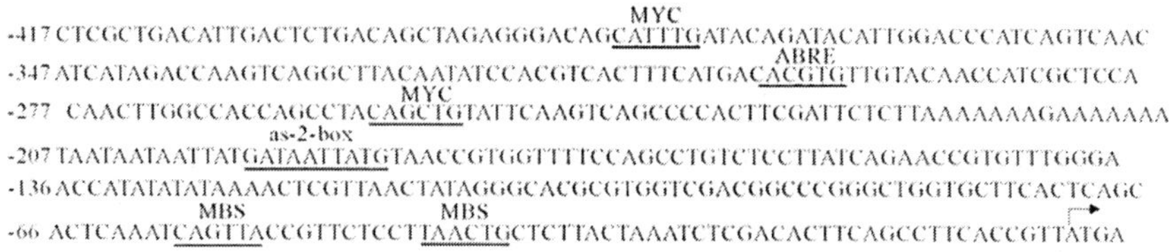

Figure 7: Distribution of putative regulatory elements in the VaCBF4 promoter from V. amurensis.

ACKNOWLEDGEMENTS

This work was supported by the China National Commonwealth (Agriculture) Science and Technology Project (No.200903044.) and the Priority Academic Program Development of Jiangsu Higher Education Institutions (PAPD).

REFERENCES

1. Willimsky, G., et al. (1992) Characterization of cspB, a Bacillus subtilis inducible cold shock gene affecting cell viability at low temperatures. Journal of Bacteriology, 174, 6326-6335.
2. Liu, J., et al. (2002) Cold signalling associated with vernalization in Arabidopsis thaliana does not involve CBF1 or abscisic acid. Physiologia Plantarum, 114, 125-134.doi:10.1034/j.1399-3054.2002.1140117.x
3. Yamaguchi-Shinozaki, K. and Shinozaki, K. (1994) A novel cis-acting element in an Arabidopsis gene is involved in responsiveness to drought, low-temperature, or high-salt stress. The Plant Cell Online, 6, 251-264. doi:10.2307/3869643
4. Baker, S.S., Wilhelm, K.S. and Thomashow, M.F. (1994) The 5′-region of Arabidopsis thaliana cor15a has cis-acting elements that confer cold-, droughtand ABA-regulated gene expression. Plant Molecular Biology, 24, 701- 713. doi:10.1007/BF00029852
5. Jiang, C., Iu, B. and Singh, J. (1996) Requirement of a CCGAC cis-acting element for cold induction of the BN115 gene from winter Brassica napus. Plant Molecular Biology, 30, 679-684. doi:10.1007/BF00049344

6. Chinnusamy, V., Zhu, J.K. and Sunkar, R. (2010) Gene regulation during cold stress acclimation in plants. Methods in Molecular Biology, 639, 39-55. doi:10.1007/978-1-60761-702-0_3

7. Medina, J.N., et al. (1999) The Arabidopsis CBF gene family is composed of three ggenes encoding AP2 domain-containing proteins whose expression is regulated by low temperature but not by abscisic acid or dehydration. Plant Physiology, 119, 463-470.doi:10.1104/pp.119.2.463

8. Haake, V., et al. (2002) Transcription factor CBF4 is a regulator of drought adaptation in Arabidopsis. Plant Physiology, 130, 639-648. doi:10.1104/pp.006478

9. Ito, Y., et al. (2006) Functional analysis of rice DREB1/ CBF-type transcription factors involved in cold-responsive gene expression in transgenic rice. Plant & Cell Physiology, 47, 141-153. doi:10.1093/pcp/pci230

10. Yang, W., et al. (2011) Dwarf apple MbDREB1 enhances plant tolerance to low temperature, drought, and salt stress via both ABA-dependent and ABA-independent pathways. Planta, 233, 219-229. doi:10.1007/s00425-010-1279-6

11. Polashock, J.J. (2010) Functional identification of a C-repeat binding factor transcriptional activator from blueberry associated with cold acclimation and freezing tolerance. Journal of the American Society for Hortcultural Science, 135.

12. Kitashiba, H., et al. (2004) Expression of a sweet cherry DREB1/ CBF ortholog in Arabidopsis confers salt and freezing tolerance. Journal of Plant Physiology, 161, 1171-1176. doi:10.1016/j.jplph.2004.04.008

13. El Kayal, W. (2006) Expression profile of CBF-like transcriptional factor genes from Eucalyptus in response to cold. Journal of Experimental Botany, 57, 2455-2469.doi:10.1093/jxb/erl019

14. Kong, Q.S. (2004) Grape in China. Chinese Agricultural Science and Technique, Beijing.

15. Xiao, H., et al. (2008) CBF4 is a unique member of the CBF transcription factor family of Vitis vinifera and Vitis riparia. Plant, Cell & Environment, 31, 1-10.

Chapter 9

Control of Light Environment: A Key Technique for High-Yield and High-Quality Vegetable Production in Protected Farmland

Xiaoyu Yang[1,2], Xiufeng Wang[1], Lijuan Wang[3], and Min Wei[1]

[1]College of Horticultural Science and Engineering, Shandong Agricultural University, Taian, China;*Corresponding Author: minwei@sdau.edu.cn

[2]School of Life Sciences, Faculty of Science, The Chinese University of Hong Kong, Hong Kong, China

[3]Department of Horticulture, Tanjin Agricultural College, Tianjin, China

ABSTRACT

Vegetable crops such as cucumber and tomato are grown widely through the world using not only field but also protected farmland. Sensitive responses of many vegetables have been widely reported to environmental conditions such as light, air temperature, relative humility, and CO_2 concentration in the past years. Among these environmental factors, light is considered to be the most important one for vegetable growth and development, especially in protected farmland. Therefore, lots of researches on effects of light environment, including light intensity, light quality, photoperiod, and light direction, on vegetable growth and development have been done in order to optimize the environmental conditions for high-yield and high-quality vegetable production in protected farmland. In this review, recent advances in light environment control for vegetable production in protected farmland have been reviewed and the prospective for the future research has been proposed as well.

INTRODUCTION

Vegetables are economically important crops and now grown most commonly through the world using not only field culture but also protected-farmland culture. Vegetable crops just like rice, wheat and maize are very sensitive to unfavorable environmental conditions, and the slight stress can influence their growth and yield. So a series of studies on the relationship between environmental factors (e.g. light, air temperature, relative humility and CO_2 concentration) and vegetable production have been done in order to optimize the environmental conditions for high-yield and high-quality vegetable production. Among these environmental factors, light is thought to be the most important one for vegetable growth and development, particularly in protected farmland. This review focused on recent advances in the control of light environment, including light intensity, light quality, photoperiod and light direction, for vegetable production in protected farmland. The perspective for future research has also been proposed.

LIGHT INTENSITY

Light intensity needed for the maximum rate of photosynthesis is quite distinct, depending on vegetable cultivars and ambient conditions. Too low light intensity can not satisfy the requirement of photosynthetic capacity and thus results in insufficient synthesis of photoassimilates, which severely influences vegetable growth, development, and yield. On the contrary, too high light intensity may cause significant decline in the photochemical activity of photosystem II or photosystem I, which is known as photoinhibition [1-4]. It usually occurs when the light-dependent reactions of photosynthesis produce ATP and NADPH in excess of that can be consumed by the reactions of dark carbon metabolism [5, 6]. Healthy leaves growing under favorable conditions can experience intense light without extensive photodamage [7]. However, when the environmental conditions do not promote carbon fixation, even weak or moderate light may become harmful [8,9]. For example, weak light can cause severe photoinhibition of cold-sensitive plants when they are exposed to low air temperature [10]. Interestingly, light requirement has been reported for cold-tolerant plants when they suffer from and acclimate clod stress recently [11,12], indicating the strong crosstalk between light and air temperature signal.

Supplemental lighting, an important light environment control technology widely used in protected farmland, can not only improve light conditions but also result in the increased air and leaf temperature in plant canopy. The increased air and leaf temperature may influence leaf photosynthesis and senescence positively or negatively, depending on seasons. As a result, the efficiency of supplemental lighting fluctuates with seasons because the environmental conditions in protected farmland are more easily controlled at low outdoor temperature in winter [13]. In order to obtain the high-quality vegetable plants under variable light conditions, the photo thermal (PTR) has been proposed [14]. The photo thermal ratio, referring to the ratio of radiant energy (mols of photosynthetic active photons per m^2) to thermal energy (degree-day), is now thought as a useful tool to identify the optimal balance between light and air temperature [15]. Therefore, finding the optimal PTR for vegetables grown in protected farmland with artificial light maybe improve the efficiency of artificial light.

LIGHT QUALITY

Light quality is thought to affect many plant physiological processes during growth and development, particularly photosynthesis and morphology. Light quality alters plant photosynthesis by the effects on the activity of photosynthetic apparatus in leaves and the effects on the expression and/or activity of the Calvin cycle enzymes [16]. Cucumbers grown under monochromatic light, including purple, blue, green, yellow and red, have reduced growth, CO_2assimilation rate and quantum yield of photosystem II electron transport as compared with plants grown under white, and these reductions are more significant in the plants under green, yellow and red. Interestingly, plants grown under purple and blue have higher stomatal conductance, total and initial Rubisco activities and higher transcriptional levels of genes encoding key enzymes in the Calvin cycle together with higher total soluble sugars, sucrose and starch contents as compared with plants grown under white, whereas in plants grown under green, yellow and red, these parameters decline. Yu and Ong [17] reported that the CO_2 assimilation rate, dark respiration, total biomass and relative growth rate of seedlings grown under monochromatic radiation were significantly lower than those of seedlings grown under broad spectrum light. Strong ultraviolet-B radiation can cause serious damage to plant photosynthesis, such as losses of both activity and content of Rubisco [18] and sedoheptulose 1,7-biphosphatase [19], inactivation of photosynthetic electron transport chain [19] and induction of stomatal closure [20]. The involvement of light quality in regulation of plant morphology through photoreceptors has been widely studied in the past decades. The phytochromes, an important photoreceptor family including phyA to phyE in Arabidopsis [21-23], can reversibly photoconvert between two conformers: the inactive red light-absorbing Pr form and the biologically active far red light-absorbing Pfr form. Pr to Pfr photoconversion takes place upon absorption of red light photons, and reversion of Pfr to Pr occurs in far red light-enriched environment and also in the dark [24]. Seedling hypocotyl elongation, which is a common phenomenon and severely decreases seedling commercial quality in seedling production, is just a well-established light-regulated response by phytochromes when seedlings are grown under continuous darkness or far redenriched ambience [25]. Recently, some new findings suggest that plant growth and development might be coregulated by photoreceptors and other endogenous factors such as hormones and temperature sensing system. For example, phytochromes and gibberellins have been reported to act in coordination to regulate multiple aspects of Arabidopsis development such as flowering and hypocotyls elongation [26]. Significant interaction between difference between day/night temperature (DIF) and end-of-day (EOD) light quality on growth, morphology, dry matter content and carbohydrate status has been observed in cucumbers [27]. Positive DIF can induce similar responses in cucumber elongation growth, day matter and carbohydrate accumulation as EOD-farred light, and further phytochrome status can interact with the responses to alternating day and night temperature. Some cases that light quality was involved in plant stress tolerance have also been reported in previous studies. For example, a low red to far red ratio light signal increases CBF (C-repeat/DRE-Binding Factor) expression in Arabidopsis, this increase being sufficient to confer freezing tolerance at temperatures higher than those required for cold acclimation [28]. All these results indicate that it may be essential to choose a suitable light quality for maximizing plant photosynthetic performance, growth and eventually their production when artificial supplemental light is used in protected farmland.

PHOTOPERIOD

Photoperiod can also influence plant growth and development, especially sex expression. Floral induction and differentiation, the most important developmental transition from vegetative growth to reproductive growth in life cycle of higher plants, can directly affects the agricultural yield through determining the time of flowering, the number of flowers and fruits, as well as the diversion of resources from vegetative growth [29,30]. Longer photoperiod can increase the daily integrated photosynthetic photon flux and more photosynthetic product can thus be synthesized and possibly release the carbohydrate competition between vegetative organs and reproductive organs. There have been many reports on effects of photoperiod on sex expression in horticultural crops such as cucumber, a model plant often used for sex expression research in flowering plants [31-33]. The number of pistillate flowers is increased under short photoperiod in monoecious cucumber [34] and in an androecious cucumber [35].

However, Cantliffe [36] reported that photoperiod had no effects on sex expression in cucumber and gherkin. Jutamanee et al. [37] reported that the influence of photoperiod on sex expression depended on the genetic background. The short day treatment promotes pistillate flower formation and suppresses staminate flower formation in monoecious cucumbers due to the increased evolution of endogenous ethylene [31], whereas the long day treatment has the reversed effect [38]. Photoperiod has no significant effects on sex expression in gynoecious cucumbers. Photoperiod can influence leaf senescence as well. Many studies have suggested that leaf senescence occurred as a consequence of shorter photoperiod, whereas extended photoperiod delayed leaf senescence [39,40]. This evidence suggests that it is necessary to take both light intensity and photoperiod into consideration as supplemental lighting is applied for crop production in protected farmland. The mean daily light integral (DLI) is thus proposed to evaluate the combined effects of light intensity and photoperiod on plant growth and development when supplemental lighting is used in practice of crops [41-43]. The DLI for optimal plant growth differs considerably between species and canopies [15]. Finding the optimal DLI for horticultural crops grown under different conditions may be an effective way to, on one hand, improve yield and, on the other hand, reduce production cost such as electricity, which is considered to be the greatest one in protected farmland with artificial light.

LIGHT DIRECTION

Light direction may be involved in regulation of plant growth and development. The ability of leaves to harvest light on different surfaces or from different directions (i.e. direct vs. diffuse rays) is determined by the interplay of structural, biochemical and physiological properties of component cell layers. Depending on azimuth, vertical leaves of plans such as Eryngium yuccifolium and Silphium terebinthinaceum may receive similar daily irradiance on the adaxial and abaxial surfaces and typically have similar photosynthetic rate when irradiated on either surface [44]. These functionally symmetrical leaves of vertical-leaf species are amphistomatous and have unifacial leaf anatomy, with palisade cells beneath the adaxial and abaxial epidermis. In contrast, the light environment of horizontal-leaf species such as Ambrosia trifida and

Solidago canadensis is highly asymmetrical, and these leaves typically have photosynthetic rates that are 30% - 50% lower when irradiated on the abaxial surface [44,45]. These functionally asymmetrical leaves of horizontal-leaf species are anatomically bifacial, with one or more layers of palisade mesophyll beneath the adaxial epidermis and spongy mesophyll beneath the abaxial epidermis. Differential responses of adaxial and abaxial leaf surfaces to light have been attributed to differences in the biochemical characteristics of stomata, palisade and spongy mesophyll cells [46-48]. These differences are similar to the characteristics of sun and shade leaves described by Boardman [49] and Bjorkman [50]. Horizontal leaves irradiated predominantly on the adaxial side during development have chloroplasts in palisade cells with sun-type properties including higher chlorophyll a/b ratio, a lower ratio of appressed to nonappressed thylakoid membranes, and higher electron transport and CO_2fixation rates than found in spongy mesophyll cells [51-53]. When horizontal leaves are irradiated on abaxial side during expansion, sun-leaf and shade-leaf properties become reversed; palisade cells are populated with shade-type chloroplasts and spongy mesophyll cells are populated with sun-type chloroplasts [52,54]. As a result, light sensitivity of leaf abaxial side expressed as stomatal conductance is significantly increased [55], indicating that adjustment of light direction might be a feasible way to increase crop canopy photosynthesis and thus improve production.

LOOK TO THE FUTURE

Now artificial light including top lighting and interlighting has been widely used in protected farmland in the region of high latitude, and as a result, yield of horticultural crops such as tomato and cucumber has increased significantly [13,15,56]. However, there are still some issues that should be further studied. For example, undoubtedly supplemental lighting significantly increases vegetable yield, but production cost ramps up at the same time because of the significantly increased electricity consumption. To smooth this problem, several possible methods can be adopted in vegetable production. The first method is to introduce some more efficient artificial light sources such as light-emitting diode (LED) to decrease electricity cost. LED is a semiconductor light source and presents many advantages including lower energy consumption, longer lifetime, improved robustness, smaller size and faster switching, over traditional light sources such as high pressure sodium lamp (HPS) and fluorescent lamps (FL), two kinds of artificial light sources now widely used in protected farmland. Some researches on effects of LED on crop growth and yield have been performed in the past several years [57], but there is still a great gap between the research and the application of LED in protected farmland. Therefore, many efforts should be made in the future.

Improving the present lighting design may improve use efficiency of artificial light and decrease electricity cost. Take cucumber as an example, previous studies have shown that interlighting increased cucumber yield and quality more significantly than top lighting due to the more even distribution of artificial light [55]. However, the further increasing of interlight proportion (Top 52% + Interlight 48%), which results in more even distribution of light in cucumber canopy, had only minor effects on the amount of yield and energy use efficiency than the treatment of Top 76% + Interlight 24% possibly due to the decrease of artificial light from vertical direction [13]. This evidence indicates that light direction might be optimized. It is well known that light absorbance is severely influenced by incident angle and reaches the maximum when the incident angle is 0°. So light use efficiency may be further improved if light incident angle to vegetable leaves could be adjust to about 0° through optimizing lighting equipment. Compared with cucumber grown with fluorescent lamps vertically installed, faster fruit growth and higher yield have been observed when cucumbers were grown with fluorescent lamps installed parallelly to cucumber leaves (unpublished data), indicating supplemental lighting based on leaf angle may be feasible.

Besides artificial lighting, improving natural light transmittance and selectivity of cover materials may be a more economic way to improve vegetable production in protected farmland in middle-latitude and low-latitude regions such as East China and Central China because solar light in these regions is enough to support the production of most vegetables. Recently, a new project named as "Study on Environment-friendly Functional Agricultural Film and Related Products", which was supported by National Science and Technology Ministry, has been started in order to develop the new-type agricultural film and related products more suitable for environmental control in the protected farmland of China and is expected to contribute greatly to the modernization of China agriculture.

ACKNOWLEDGEMENTS

This work has been supported by The National Key Technology R&D Program of China (2012BAD11B01) and China Agriculture Research System (CARS-25-D-03).

REFERENCES

1. García-Plazaola, J.I., Becerril, J.M., Hernández, A., Niinemets, Ü. and Kollist, H. (2004) Acclimation of antioxidant pools to the light environment in a natural forest canopy. New Phytologist, 163, 87-97. doi:10.1111/j.1469-8137.2004.01096.x
2. Szabo, I., Bergantino, E. and Giacometti, G.M. (2005) Light and oxygenic photosynthesis: Energy dissipation as a protection mechanism against photo-oxidation. EMBO Reports, 6, 629-634. doi:10.1038/sj.embor.7400460
3. Kreslavski, V.D., Carpentier, R., Klimov, V.V., Murata, N. and Allakhverdiev, C.I. (2007) Molecular mechanism of stress resistance of the photosynthetic apparatus. Biochemistry (Moscow) Supplement Series A: Membrane and Cell Biology, 1, 185-207.doi:10.1134/S1990747807030014
4. Murata, N., Takahashi, S., Nishiyama, Y. and Allakhverdiev, S.I. (2007) Photoinhibition of photosystem II under environmental stress. Biochimica and Biophysica Acta, 1767, 414-421. doi:10.1016/j.bbabio.2006.11.019
5. Demmig-Adams, B. and Adams, W.W. (2000) Photosynthesis: Harvesting sunlight safely. Nature, 403, 371-374. doi:10.1038/35000315
6. Li, Y.Y., Sperryb, J.S. and Shao, M.G. (2009) Hydraulic conductance and vulnerability to cavitation in corn (Zea mays L.) hybrids of differing drought resistance. Environmental and Experimental Botany, 66, 341-346. doi:10.1016/j.envexpbot.2009.02.001
7. Govindacharya, S., Bukhovab, N.G., Jolya, D. and Carpentiera, R. (2004) Photosystem II inhibition by moderate light under low

temperature in intact leaves of chilling-sensitive and -tolerant plants. Physiologia Plantarum, 121, 322-333. doi:10.1111/j.0031-9317.2004.00305.x

8. Kudoh, H. and Sonoike, K. (2002) Irreversible damage to photosystem I by chilling in the light: Cause of the degradation of chlorophyll after returning to normal growth temperature. Planta, 215, 541-548. doi:10.1007/s00425-002-0790-9

9. Gerotto, C., Alboresi, A., Giacometti, G.M., Bassi, R. and Morosinotto, T. (2011) Role of PSBS and LHCSR in Physcomitrella patens acclimation to high light and low temperature. Plant, Cell and Environment, 34, 922-932. doi:10.1111/j.1365-3040.2011.02294.x

10. Krause, G.H. (1994) Photoinhibition induced by low temperatures. In: Baker, N.R. and Bowyer, J.R., Eds., Photoinhibition of Photosynthesis: From Molecular Mechanisms to the Field, BIOS Scientific Publishers, Oxford.

11. Kim, H.J., Kim Y.K., Park, J.Y. and Kim, J. (2002) Light signaling mediated by phytochrome plays an important role in cold-induced gene expression through the C-repeat/dehydration responsive element (C/DRE) in Arabidopsis thaliana. Plant Journal, 29, 693-704. doi:10.1046/j.1365-313X.2002.01249.x

12. Catalá, R., Medina, J. and Salinas, J. (2011) Integration of low temperature and light signaling during cold acclimation response in Arabidopsis. Proceedings of the National Academy of Sciences of the United States of America, 108, 16475-16480.doi:10.1073/pnas.1107161108

13. Hovi-Pekkanen, T. and Tahvonen, R. (2008) Effects of interlighting on yield and external fruit quality in yearround cultivated cucumber. Scientia Horticulturae, 116, 152-161.doi:10.1016/j.scienta.2007.11.010

14. Liu, B. and Heins, R.D. (2002) Photothermal ratio affects plant quality in "Freedom" poinsettia. Journal of America Society for Horticultural Science, 127, 20-26.

15. Moe, R., Grimstad, S.O. and Gislerød, H.R. (2006) The use of artificial light in year round production of greenhouse crops in Norvey. Acta Horticulturae, 711, 35-42.

16. Wang, H., Gu, M., Gui, J.X., Shi, K., Zhou, Y.H. and Yu, J.Q. (2009) Effects of light quality on CO_2 assimilation, chlorophyll-fluorescence quenching, expression of Calvin cycle genes and carbohydrate accumulation in Cucumis sativus. Journal of Photochemistry and Photobiology B: Biology, 96, 30-37. doi:10.1016/j.jphotobiol.2009.03.010

17. Yu, H. and Ong, B.L. (2003) Effect of radiation quality on growth and photosynthesis of Acacia mangium seedlings. Photosynthetica, 41, 349-355.doi:10.1023/B:PHOT.0000015458.11643.b2

18. Allen, D.J., McKee, I.F., Farage, P.K. and Baker, M.R. (1997) Analysis of limitations to CO_2assimilation on exposure of leaves of two Brassica napus cultivars to UV-B. Plant, Cell and Environment, 20, 633-640. doi:10.1111/j.1365-3040.1997.00093.x

19. Allen, D.J., Nogués, S. and Baker, N.R. (1998) Ozone depletion and increased UV-B radiation: Is there a real threat to photosynthesis? Journal of Experimental Botany, 49, 1775-1788.

20. Nogués, S. and Baker, N.R. (1995) Evaluation of the role of damage to photosystem II in the inhibition of CO_2 assimilation in pea leaves on exposure to UV-B. Plant, Cell and Environment, 18, 781-787. doi:10.1111/j.1365-3040.1995.tb00581.x

21. Rockwell, N.C., Su, Y.S. and Lagarias, J.C. (2006) Phytochrome structure and signaling mechanisms. Annual Review of Plant Biology, 57, 837-858.doi:10.1146/annurev.arplant.56.032604.144208

22. Schafer, E. and Nagy, F. (2006) Photomorphogenesis in plants and bacteria. Springer, Dordrecht. doi:10.1007/1-4020-3811-9

23. Quail, P.H. (2010) Phytochromes. Current Biology, 20, R504-R507.doi:10.1016/j.cub.2010.04.014

24. Leivar, P., Montec, E., Megan, M., Cohn, M.M. and Quail, P.H. (2012) Phytochrome signaling in green arabidopsis seedlings: Impact assessment of a mutually negative phyB-PIF feedback loop. Molecular Plant, 5, 208-223. doi:10.1093/mp/sss031

25. Soy, J., Leivar, P., González-Schain, N., Sentandreu, M., Prat, S., Quail, P.H. and Monte, E. (2012) Phytochromeimposed oscillations in PIF3 protein abundance regulate hypocotyl growth under diurnal light/dark conditions in Arabidopsis. Plant Journal, 71, 390-401.

26. Facella, P., Daddiego, L., Giuliano, G. and Perrotta, G. (2012) Gibberellin and auxin influence the diurnal transcription pattern of photoreceptor genes via CRY1a in tomato. PLoS ONE, 7, e30121. doi:10.1371/journal.pone.0030121
27. Xiong, J.Q., Patil, G.G., Moe, R. and Torre, S. (2011) Effects of diurnal temperature alternations and light quality on growth, morphogenesis and carbohydrate content of Cucumis sativus L. Scientia Horticulturae, 128, 54-60. doi:10.1016/j.scienta.2010.12.013
28. Franklin, K.A. and Whitelam, G.C. (2007) Light-quality regulation of freezing tolerance in Arabidopsis thaliana. Nature Genetics, 39, 1410-1413. doi:10.1038/ng.2007.3
29. Terefe, D. and Tatlioglu, T. (2005) Isolation of a partial sequence of a putative nucleotide sugar epimerase, which may involve in stamen development in cucumber (Cucumis sativus L.). Theoretical and Applied Genomics, 111, 1300-1307. doi:10.1007/s00122-005-0058-4
30. Chen, H.M., Tian, Y., Lu, X.Y. and Liu, X.H. (2011) The inheritance of two novel subgynoecious genes in cucumber (Cucumis sativus L.). Scientia Horticulturae, 127, 464-467. doi:10.1016/j.scienta.2010.11.004
31. Yamasaki, S., Fujii, N. and Takahashi, H. (2003) Photoperiodic regulation of CS-ACS2, CS-ACS4 and CS-ERS gene expression contributes to the femaleness of cucumber flowers through diurnal ethylene production under short-day conditions. Plant, Cell and Environment, 26, 537-546. doi:10.1046/j.1365-3040.2003.00984.x
32. Tanurdzic, M. and Banks, J.A. (2004) Sex-determining mechanisms in land plants. The Plant Cell, 16, S61-S71. doi:10.1105/tpc.016667
33. Miao, M.M., Yang, X.G., Han, X.S. and Wang K.S. (2011) Sugar signaling is involved in the sex expression response of monoecious cucumber to low temperature. Journal of Experimental Botany, 62, 797-804. doi:10.1093/jxb/erq315
34. Astmon, D. and Galun, E. (1962) Physiology of sex in Cucumis sativus (L.) leaf age patterns and sexual differentiation of floral buds. Annals of Botany, 26, 137-146.

35. Rudich, J., Baker, L.R., Scott, J.W. and Sell, H.M. (1976) Phenotypic stability and ethylene evolution in androecious cucumber. Journal of the American Society for Horticultural Science, 101, 48-51.

36. Cantliffe, D.J. (1981) Alteration of sex expression in cucumber due to changes in temperature, light intensity, and photoperiod. Journal of the American Society for Horticultural Science, 106, 133-136.

37. Jutamanee, K., Saito, T. and Subhadrabandhu, S. (1994) Control of sex expression in cucumber by photoperiod, defoliation, and plant growth regulators. Kasetsart Journal (Natural Science), 28, 626-631.

38. Wu, T., Qin, Z.W., Zhou, X.Y., Feng, Z. and Du, Y.L. (2010) Transcriptome profile analysis of floral sex determination in cucumber. Journal of Plant Physiology, 167, 905-913. doi:10.1016/j.jplph.2010.02.004

39. Rosenthal, S.I. and Camm, E.L. (1996) Effects of air temperature, photoperiod and leaf age on foliar senescence of western larch (Larix occidentalis Nutt.) in environmentally controlled chambers. Plant, Cell and Environment, 19, 1057-1065. doi:10.1111/j.1365-3040.1996.tb00212.x

40. Zhao, H., Li, Y., Duan, B., Korpelainen, H. and Li, C. (2009) Sex-related adaptive responses of Populus cathayana to photoperiod transitions. Plant, Cell and Environment, 32, 1401-1411. doi:10.1111/j.1365-3040.2009.02007.x

41. Moccaldi, L.A. and Runkle, E.S. (2007) Modeling the effects of temperature and photosynthetic daily light integral on growth and flowering of Salvia splendens and Tagetes patula. Journal of the American Society for Horticultural Science, 132, 283-288.

42. Oh, W., Cheon, I.H., Kim, K.S. and Runkle, E.S. (2009) Photosynthetic daily light integral influences flowering time and crop characteristics of Cyclamen persicum. HortScience, 44, 341-344.

43. Garland, K.F., Burnett, S.E., Stack, L.B. and Zhang, D.L. (2010) Minimum daily light integral for growing highquality coleus. HortTechnology, 20, 929-933.

44. DeLucia, E.H., Shenoi, H.D., Naidu, S.L. and Day, T.A. (1991) Photosynthetic symmetry of sun and shade leaves of different orientations. Oecologia, 87, 51-57.doi:10.1007/BF00323779

45. Terashima I. (1989) Productive structure of a leaf. In: Briggs, W.R., Ed., Plant Biology, Photosynthesis, 8, Alan R. Liss Inc., New York.

46. Poulson, M.E. and DeLueia, E.H. (1993) Photosynthetic and structural acclimation to light direction in vertical leaves of Silphium terebinthiaaceum. Oecologia, 95, 393- 400. doi:10.1007/BF00320994

47. Soares, A.S., Driscoll, S.P., Olmoe, E., Arrabaca, M.C. and Foyer, C.H. (2008) Adaxial/abaxial specification in the regulation of photosynthetic CO_2 assimilation with respect to light orientation and growth with CO_2 enrichment in Paspalum dilatatum leaves. New Phytologist, 177, 186-198.

48. Wang, Y., Noguchi, K. and Terashima, I. (2008) Distinct light responses of the adaxial and abaxial stomata in intact leaves of Helianthus annuus L. Plant, Cell and Environment, 31, 1307-1316. doi:10.1111/j.1365-3040.2008.01843.x

49. Boardman, N.K. (1977) Comparative photosynthesis of sun and shade plants. Annual Review of Plant Physiology, 28, 355-377. doi:10.1146/annurev.pp.28.060177.002035

50. Bjorkman, O. (1981) Responses to different quantum flux densities. In: Lange, O.L., Nobel, P.S., Osmond, C.B. and Zeigler, H., Eds., Physiological Plant Ecology I., Ency in Plant Physiology, NS, 12A, Springer, New York. doi:10.1007/978-3-642-68090-8_4

51. Schreiber, U., Fink, R. and Vidaver, W. (1977) Fluorescence induction in whole leaves: Differentiation between the two teaf sides and adaptation to different light regimes. Planta, 133, 121-129. doi:10.1007/BF00391909

52. Terashima, I. and Inoue, Y. (1984) Comparative photosynthetic properties of palisade tissue chloroplasts and spongy tissue chloroplasts of Camellia japonica L.: Functional adjustment of the photosynthetic apparatus to light environment within a leaf. Plant and Cell Physiology, 25, 555-563.

53. Terashima, I. and Takenaka, A. (1986) Organization of photosynthetic system of dorsiventral leaves as adapted to the irradiation from the adaxial side. In: Marcell, R., Clijsters, H. and Van Poucke, M., Eds., Biological Control of Photosynthesis. Martinus Nijhoff, Dordrecht. doi:10.1007/978-94-009-4384-1_20

54. Terashima, I. (1986) Dorsiventrality in photosynthetic light response curves of a leaf. Journal of Experimental Botany, 37, 399-405. doi:10.1093/jxb/37.3.399

55. Wang, Y., Noguchi, K. and Terashima, I. (2011) Photosynthesis-dependent and -independent responses of stomata to blue, red and green monochromatic light: Differences between the normally oriented and inverted leaves of sunflower. Plant and Cell Physiology, 52, 479-489. doi:10.1093/pcp/pcr005

56. Hovi, T., Näkkilä, J. and Tahvonen, R. (2004) Interlighting improves production of year-round cucumber. Scientia Horticulturae, 102, 283-294.doi:10.1016/j.scienta.2004.04.003

57. Trouwborst, G., Oosterkamp, J., Hogewoning, S.W., Harbinson, J. and van Ieperen, W. (2010) The responses of light interception, photosynthesis and fruit yield of cucumber to LED-lighting within the canopy. Physiologia Plantarum, 138, 289-300.doi:10.1111/j.1399-3054.2009.01333.x

Citations

CHAPTER 1

Edward Martey, Prince M Etwire, Alexander N Wiredu, and Wilson Dogbe, Factors influencing willingness to participate in multi-stakeholder platform by smallholder farmers in Northern Ghana: implication for research and development, doi:10.1186/s40100-014-0011-4.

CHAPTER 2

Carol Rolando, Robyn Gaskin, David Horgan, Nari Williams, and Martin K-F Bader, The Use of Adjuvants to Improve Uptake of Phosphorous Acid Applied to Pinus Radiata Needles for Control of Foliarphytophthora Diseases, doi:10.1186/s40490-014-0008-5.

CHAPTER 3

Lars Koschke, Carsten Lorz, Christine Fürst, Tobias Lehmann, and Franz Makeschin, Assessing Hydrological And Provisioning Ecosystem Services in a Case Study in Western Central Brazil, doi:10.1186/2192-1709-3-2.

CHAPTER 4

Christopher J Short and Janet Dwyer, Reconciling pastoral agriculture and nature conservation: developing a co-management approach in the English uplands, doi:10.1186/2041-7136-2-13.

CHAPTER 5

Reuben P Keller, Juergen Geist, Jonathan M Jeschke, and Ingolf Kühn, Invasive species in Europe: ecology, status, and policy, doi:10.1186/2190-4715-23-23.

CHAPTER 6

Serge D Elie, Pastoralism in Soqotra: external entanglements and communal mutations, doi: 10.1186/s13570-014-0016-3.

CHAPTER 7

Joseph G Dubouzet, Lloyd Donaldson, Michael A Black, Les McNoe,

Vincent Liu, and Gareth Lloyd-Jones, Heterologous Hybridisation to a Pinus Microarray: Profiling of Gene Expression in Pinus Radiata Saplings Exposed to Ethephon, doi:10.1186/s40490-014-0021-8.

CHAPTER 8

Dong, C. , Zhang, M. , Yu, Z. , Ren, J. , Qin, Y. , Wang, B. , Xiao, L. , Zhang, Z., and Tao, J. (2013) Isolation and expression analysis of CBF4 from *Vitis amurensis* associated with stress. *Agricultural Sciences*, 4, 224-229. doi:10.4236/as.2013.45032.

CHAPTER 9

Yang, X. , Wang, X. , Wang, L. and Wei, M. (2012) Control of light environment: A key technique for high-yield and high-quality vegetable production in protected farmland. *Agricultural Sciences*, 3, 923-928. doi:10.4236/as.2012.37112.

Index